Modelling
Marine
Processes

Modelling Marine Processes

Phil Dyke
University of Plymouth

PRENTICE HALL

*London   New York   Toronto   Sydney   Tokyo   Singapore
Madrid   Mexico City   Munich*

First published 1996 by
Prentice Hall Europe
Campus 400, Maylands Avenue
Hemel Hempstead
Hertfordshire, HP2 7EZ
A division of
Simon & Schuster International Group

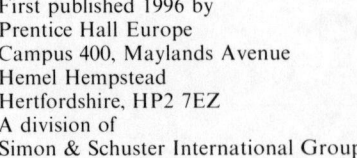

© Prentice Hall 1996

All rights reserved. No part of this publication may be reproduced, stored in a retrieval system, or transmitted, in any form, or by any means, electronic, mechanical, photocopying, recording or otherwise, without prior permission, in writing, from the publisher.
For permission within the United States of America contact Prentice Hall Inc., Englewood Cliffs, NJ 07632

Typeset in 10/12pt Times
by PPS Limited, London Road, Amesbury, Wiltshire

**Printed and bound in Great Britain by
T. J. Press (Padstow) Ltd.**

Library of Congress Cataloging-in-Publication Data

Available from the publisher.

British Library Cataloguing in Publication Data

A catalogue record for this book is available from the British Library

ISBN 0-13-098120-9

1 2 3 4 5 00 99 98 97 96

To Eleanor, my younger daughter

Contents

Preface   xi

1 Introduction to modelling   **1**
  1.1 Preliminary remarks   1
  1.2 The modelling process   2
  1.3 The systems approach   4

2 Dynamic balances in oceanography   **6**
  2.1 Introduction   6
  2.2 Fundamental balances   6
  2.3 Other balances   12
  2.4 Internal balances   13
  2.5 Other dimensional quantities   14
  2.6 Examples of dimensional analysis   15
    2.6.1 The Gulf Stream   15
    2.6.2 Parametrising bottom friction   16

3 Numerical methods   **18**
  3.1 Introduction   18
  3.2 Numerical methods in oceanography   18
  3.3 Errors   21
  3.4 Examples using numerical methods   23
    3.4.1 A semi-implicit finite difference scheme   23
    3.4.2 A finite element method for tides in the western Atlantic   25
    3.4.3 Finite difference grids for ocean models   27

4 Boundary conditions and validation   **29**
  4.1 Introduction   29
  4.2 The sea surface boundary   31
  4.3 The sea bed boundary   32
  4.4 Coastal boundary conditions   34

	4.5	Open boundary conditions	35
	4.6	Model validation	37
	4.7	Examples of representing boundaries	38
		4.7.1 Finite difference schemes	38
		4.7.2 Finite element schemes	40

5 Large-scale ocean dynamics — 42

- 5.1 Introduction — 42
- 5.2 A perspective on global ocean modelling — 43
- 5.3 Steady ocean circulation models — 45
- 5.4 The Gulf Stream and other large-scale phenomena — 48
- 5.5 Modelling large-scale phenomena — 50
- 5.6 Modern ocean modelling — 52
- 5.7 Two detailed modelling examples — 53
  - 5.7.1 The Semtner model — 53
  - 5.7.2 *El Niño* — 56

6 Continental shelf sea modelling — 60

- 6.1 Introduction — 60
- 6.2 Tidal modelling — 60
- 6.3 Wind-driven circulation — 65
- 6.4 Density-driven and wave-driven flows — 66
- 6.5 Two modelling examples — 70
  - 6.5.1 The North Sea — 71
  - 6.5.2 The Firth of Forth and the Forth Estuary — 73

7 Diffusion modelling — 77

- 7.1 Introduction — 77
- 7.2 Modelling the process of diffusion — 78
- 7.3 Some specific diffusion models — 79
- 7.4 Diffusion models that track particles — 81
- 7.5 Modelling diffusion in the North Sea — 85

8 Ecosystem modelling — 88

- 8.1 Introduction — 88
- 8.2 Predator–prey ecosystems — 89
- 8.3 Modelling a real marine ecosystem — 91
- 8.4 Other ecological models — 97

9 Modelling using a programmed text approach — 103

- 9.1 Introduction — 103
- 9.2 Statistics — 103
- 9.3 Modelling in action — 118
- 9.4 Exercises — 128

Appendix A	Mathematical modelling in oceanography	133
	A.1 Dimensionless numbers	133
	A.2 An equatorial model	134
	A.3 Modelling *El Niño*	137
Appendix B	The χ^2 distribution	139
Appendix C	Commercially available software	141
	C.1 Commissioned software	141
	C.2 The ECoS system	143

Answers to exercises   145

References   146

Index   149

Preface

At the outset, it is essential to realise that this is not a mathematics text, even though it is written by an applied mathematician.

In oceanography as in most branches of applied science, the arrival of the desktop microcomputer has revolutionised what is possible in terms of modelling. No longer is it necessary to be competent in programming in a high-level computer language in order to be able to do modelling. Sophisticated software is available which, after a few keystrokes of selection from a pop-up menu, gives complicated multicoloured views often in movie style of flows in your choice of sea area. The key point here is the potential lack of expertise of the user. Literally anybody can make choices from the menu and get the glamorous output. Modern easy-to-use software is therefore open, if not to abuse then certainly to misinterpretation. Do we therefore go back to the days of the 1970s when in order to do any modelling at all required long experience in writing computer programs, usually in FORTRAN, and a user number on a mainframe computer? Even if access these days is via a terminal instead of stacks of cards and is blissfully free of that gobbledegook found at the start of old computer code and which went under the name of a command language. The answer to the question must be of course not! We must move with the times but guard against the misuse of sophisticated software. This is where, it is hoped, the present text is most useful. One of its main aims is to enable those interested in marine modelling – including the environmental scientist, the environmental engineer, the ocean scientist, the physicist, chemist or biologist with marine interests – to be able to use and to review with informed criticism a model which they have not and indeed could not have built, but have perhaps bought, have had commissioned, or have read details about in the literature.

This book has been developed out of courses the author has given to both undergraduate ocean scientists and postgraduate marine scientists at Plymouth over a period of years. The background of these students is very broad: on the one hand, there are mature students who are very keen to learn, but who last studied 20 or so years before and whose mathematical skills and, more importantly, confidence are lacking; then there are students with a good 'A' level in mathematics together with a strong interest in matters concerned with the sea who by and large can cope with technical material after a little practice. This vast range presents a challenge to the teacher, particularly when the class size exceeds 50. It has been found that the approach used in this book remains user-friendly to students of either extreme. The lack of mathematics proves to be a gentle way to introduce modelling to those who lack the technical background, whilst it is also useful for the more mathematically sophisticated to see a different approach even if they are smiling smugly to themselves and whispering 'I think I know what's going on here'. More mathematically explicit material is given in Appendix A for completeness. In Appendix C a short account of the situation regarding available software is given, including an example of a system shell.

The book begins with an introduction to modelling in general, and to dimensional analysis as applied to physical oceanography in particular. There follows a chapter on those numerical methods that have been used by marine modellers. It is perhaps a surprise that this chapter is virtually devoid of mathematics. The methods are described rather than given in mathematical detail, although the examples are mathematically explicit. Chapter 4 is concerned with the all-important matter of boundary conditions. These occur in the form of coasts, the sea surface, the sea bed, as well as the start conditions and conditions to be applied on an open boundary which is a boundary to the model that does not correspond to any physical feature of the sea itself. Again, there is not much in the way of mathematics in this chapter. The emphasis as always is on telling the reader what is going on. In Chapters 5, 6, 7, and 8 various aspects of marine science are covered and examples of models given. Chapter 8, which is about ecosystem modelling, deserves a special mention in view of the absence of this topic from many contemporary oceanographic texts. There are, I suppose, two principal reasons for this absence. First, there is no equivalent to Newton's laws in biology, so the models are less well founded in universally accepted scientific laws. Second, modelling ecosystems is still a rather new field of research and models still lack the sophistication of the physical models in which they are often embedded. In the last ten or so years, however, models have appeared that do stand up to rigorous validation. The most uncomfortable feature of ecosystem models remains their over-reliance on poorly determined parameters.

Chapter 9 is different. What has been attempted here is a programmed learning format for tutorial-style problems. When reading this chapter be prepared to do some calculating! The best way to understand modelling is to have a try at doing some. Again, the examples used are those that have proved most successful in the classroom, and the format used has also proved highly successful, albeit for students of other more traditional areas of applied science and engineering.

Finally, a word about what is absent. I have decided to exclude from this text those areas of sea surface modelling that normally fall under the scope of the civil or hydraulic engineer. There is no discussion therefore of wave theory, design currents for the offshore industry, or extreme currents as might be caused by the unfortunate combination of tide, wind-induced flow and other residual flow. If this text runs to a second edition, these are candidates for inclusion then!

It is a pleasure to thank Bryan Johns and John Johnson for introducing various aspects of oceanography to a mathematician all those years ago. I thank Cliff Johnston and Martin Wilkinson who first asked me to deliver physical oceanography lectures to students with no mathematical background. Thanks go to all my students over the years for the help that they have given me in the form of feedback, even that which at the time would have been hard to classify as help! Thanks also to all those publishers of books and journals who have given me permission to reproduce diagrams from their publications. Finally, thanks to David Huntley who, though he may not remember it, actually suggested the very apt title for this book.

Phil Dyke

Chapter One

Introduction to modelling

1.1 Preliminary remarks

To the majority, the word modelling either means something to do with photography or, if they have any scientific background, the building of scaled-down replicas that ought to mimic real life situations. In this latter category one thinks of civil engineering consultants building models of harbours with attendant breakwaters and jetties, and then subjecting them to a particular wave climate. The way this is done is to build a physical model, usually in a large area reminiscent of an aircraft hanger. In this model, the area of coast or river or estuary (whatever) is built from materials such as concrete and builders' rubble. Of course there is a scale, perhaps 1:20 or even larger, which needs to be considered when examining results. If waves are of interest, then there has to be a paddle mechanism included to generate them. Exactly how the scale factors can be calculated is the subject of Chapter 2, but suffice it to say that measurements in terms of wave heights, current strengths, forces on piers, etc., are taken, and from them, after application of the appropriate scaling factor, an estimate of the real life wave, current or force can be made. Up to 30 years ago virtually all modelling in coastal engineering was done in this fashion in special laboratories, for example Hydraulics Research in the UK, or Delft Hydraulics in the Netherlands. The outcome may be the identification of a particularly sensitive area where erosion or flooding is likely to occur, or perhaps a strain gauge or two could give information on likely strains and hence stresses of appropriately scaled piers and jetties. The use of scale factors then gives the real stresses and strains. This is but one case where physical modelling plays a part. Throughout engineering and applied science similar modelling still plays an important role. Space exploration (the design of rockets and satellites), and the design of bridges are other areas that spring to mind.

However, this is not what is meant by modelling in this text (although there are parallels, as explained in Chapter 2). Mathematics is often thought of as a very well defined set of axioms and techniques that give precise answers to well defined problems. Such a discipline, as it stands, is not well suited to describe a practical science such as oceanography. The very mathematical papers which purport to be oceanography usually suffer from a tenuous relationship with reality. On the other hand, at the other end of the spectrum, there are very simple mathematical expressions that contain oceanographic truths. This blending of mathematics and oceanographic processes is the field in which the modelling being discussed here lies. In modern times, this modelling is greatly assisted by computing; it is probably safe to assume that the modelling referred to here occurs on a computer. Although a computer is by no means essential to do this kind of modelling, most practical modelling makes use of the power, speed and graphical display facilities of microcomputers, minicomputers, mainframes or perhaps even supercomputers.

Oceanography, which can be assumed to be synonymous with marine science, is a science which has grown through painstaking observation and progressed through scientists making judgements and deductions from these observations. There are several distinctive features that, although not peculiar to oceanography, meteorology and Earth sciences in general, help make it particularly amenable to the relatively new art of modelling. (Yes, although there is some scientific rigour in modelling it still remains very much an art.) First, oceanography as an applied science has to incorporate aspects of physics, chemistry and biology. Indeed, it may be argued that the sea provides an ideal vehicle for the study of some (but certainly not all) of the fundamentals of the basic sciences. In order to understand some of the processes that occur in the sea, it is therefore necessary to simplify some aspects and ignore others. This is what occurs in modelling. Second, there is a very important aspect to modelling called validation. In most sciences and engineering validation means trying out the model and comparing it with the real situation. In oceanography the whole history of the science, from the accumulated wisdom of fishermen through the voyages of discovery to modern day scientific expeditions is centred around observation. This provides in some respects an ideal scenario for validation, except that conditions are not controlled, so there is no control over input.

1.2 The modelling process

Many books on modelling begin by describing an idealised modelling process, often using as a vehicle, some simple problem. It is perhaps difficult for the practical oceanographer to be entirely convinced of the merits of modelling because of the element of trial and error involved. However, it is precisely this heuristic aspect of modelling that enables it to be successful in its mimicry of real life.

As mentioned earlier, an important element of modelling which has led to its recent popularity is the ready availability of cheap but powerful computers. In the whole of the 1960s and 1970s and part of the 1980s, in order to use computers it was necessary to learn how to program them, and in order to program them it was necessary to become familiar with a high-level programming language such as ALGOL (in the early days), FORTRAN, PASCAL or C. The details of the programming, in turn, demand equally detailed knowledge of the mathematics that underlie the model itself. By their very nature, these programs utilising as they do powerful computers are complicated and are based on sophisticated rather than simple mathematics. A requirement for those that were involved in marine modelling was therefore some

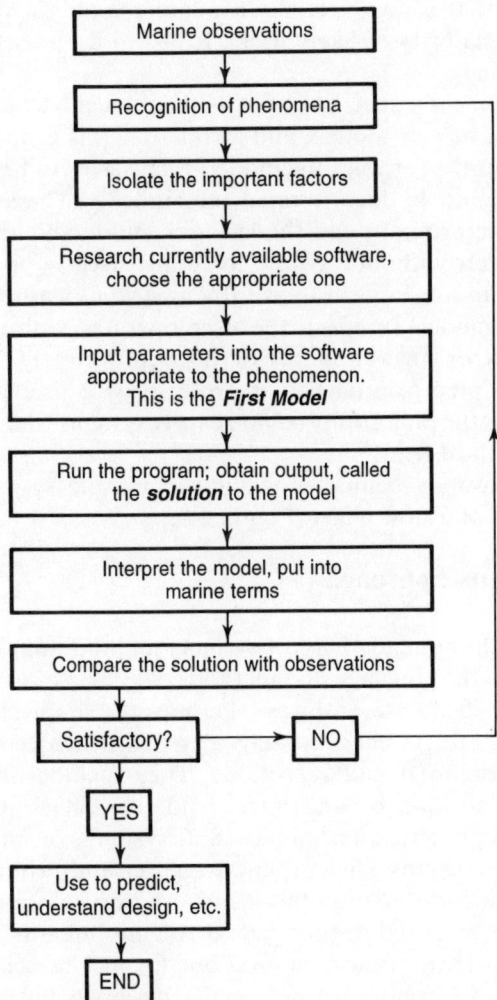

Figure 1.1 A marine modelling process.

knowledge of mathematics including the calculus that is used to describe the dynamic balances in a fluid and the transport of heat and salt, and the methods of discretising which comprise that branch of mathematics called numerical methods. Much of this kind of mathematical modelling is still of course going on, but it is no longer, I believe, mandatory to be as close to the mathematics. Many marine scientists are concerned with models on computers because they wish to answer engineering or environmental impact questions, but they lack the mathematical background to formulate and program models on computers. Software (the modern name for a

computer program that is commercially available) is only a successful product if it is accessible to the majority of likely users. It has to do a useful job too, of course. It is widely recognised
that not enough marine scientists have the mathematical background to comprehend the details of today's marine models, and fortunately this is no longer necessary. The very computer power that enables the models themselves to be complex also enables so-called 'front ends' to be incorporated into models. These front ends act as an interface between the program and the lay user and enable the lay user to use the program constructively without the need to get involved in programming itself. One common method is to use English to ask the user what features he or she wishes to incorporate into a model. In effect, the user operates with a selection of menus, choosing from a set or answering yes or no appropriately. This way, a complex program can be adapted to a particular problem by a user with no knowledge of programming. This is the philosophy behind expert systems which are now widespread, especially in the medical field.

The general philosophy behind modelling in marine systems can be expressed succinctly by means of a flow chart (Figure 1.1).

1.3   The systems approach

Some mention should be made here of systems methodology. Many engineers and applied scientists use the language and methods of systems successfully in their work. Systems may be classified in several ways. Perhaps the simplest partition is into hard and soft systems. A hard system is, loosely, a system which acts independently of the observer, is well structured and repeatable. They include all the traditional well controlled engineering that, by and large, lend themselves to exact mathematical solutions or at least numerical techniques. Soft systems, on the other hand, include ill-posed problems, problems where opinions matter and problems that do not have a single solution, but have many solutions or even no solutions at all. Soft systems are more usually encountered in problems involving humans. An alternative classification of models, perhaps a more natural one for marine scientists, is into natural systems (e.g. biological organisms), artificially designed but physical systems (e.g. engineering devices), and artificially designed but abstract systems (e.g. economic models and models involving scheduling). Marine science modelling as outlined in Section 1.1 falls naturally into the first of these categories. However, there is a very substantial difference between modelling the physics of the sea and modelling the biology of the sea. In the former, there are well established laws that can be invoked, whereas in biology the laws are less universal and are developed to be consistent with data rather than from a fundamental understanding of processes. Having said this, in the language of systems it remains true that most marine modelling can be classified as hard systems methodology. It is only very recently that marine scientists have brushed with problems that verge on being soft in the language of systems. This is because the more complicated the model, the less precise is the solution technique, and the more debatable are the reasons behind a particular outcome.

It is perhaps tempting to think of modelling as a well established set of principles and routines. However, this is only true to a certain extent, particularly for simple problems. As more variables are included in a model of a marine system, there are more parameters which are to some extent at the behest of the modeller. These might be the extent to which the rate of change (growth) of a particular organism is dependent upon another in a biological model, or it may be how a particular variable such as salinity dilutes with time in the presence of a particular current. This dilution is governed (usually) by a single parameter (see Chapter 7 for more details). When the numbers of such parameters increase, then so must the level of uncertainty associated with outcomes of the model. This is not a criticism of complex models; it is merely a consequence of having so many adjustable parameters. (Note here the distinction between 'variable' and 'parameter'. *Variables* are quantities whose behaviour we wish to model; *parameters* are quantities that can be estimated either directly or indirectly, and occur as a direct consequence of modelling assumptions. Parameters are not natural quantities, whereas variables are.) Of course if there is a perceived fault in the outcome of a model, it is usually a matter of conjecture whether any fault lies in what has been left out of the model, the method the software uses, or the interpretation of the output. Perhaps the observations themselves are in error! It is usually only experience that comes to our aid. However, it is precisely the principal aim of this text to give the marine scientists this experience through the eyes of the author.

Chapter Two

Dynamic balances in oceanography

2.1 Introduction

Much of this chapter discusses *typical* lengths, *typical* times, *typical* speeds, etc. It is by no means obvious what is meant by this, so a few words of introduction are in order before getting down to specific details. Many variables can be expressed in terms of the so-called fundamental quantities: time (measured in seconds), length (measured in metres) and mass (measured in kilograms). Some of these quantities are speed (length divided by time), density (mass divided by volume where volume is of course length × length × length) and pressure (mass divided by (length times (time)^2)). In a given situation, most users of models wish to concentrate on phenomena which have a specific limited range of any of the fundamental quantities, or variables derived from them. For example, a tidal modeller would not be interested in small time scales of a few seconds or large time scales of many centuries, but something in between; an estuarial modeller would not be concerned with length scales of hundreds or thousands of kilometres, but something much less. Most phenomena in marine science are scale specific, and the art of successful modelling is often linked with the ability to screen out the unwanted in order to focus on what is desired. Dimensional analysis forms an essential element in this process of simplification.

2.2 Fundamental balances

It has now been over 300 years since Sir Isaac Newton formulated his famous view of mechanics. Despite the recent advances in physics which have led to new theories of the very large and fast (general relativity, ∼ 1915) and the very small and also fast (quantum mechanics, ∼ 1926), the motion of virtually all known objects closely follows the laws set down by Newton all those years ago. Newton's laws of mechanics certainly apply on all the length and time scales appropriate to marine science. The adaptation of Newton's second law to fluids (which of course include both air and sea water) occurred in the eighteenth century through the work of Euler and (Daniel) Bernoulli. However, their reformulation does not change the basic equation, which Newton formulated as

$$\text{force} = \text{mass} \times \text{acceleration},$$

and which in fluids reads

$$\text{force per unit mass} = \text{acceleration}$$

(Bernoulli's contribution was to convert the equation of fluid motion to the form of an equation which expresses conservation of mechanical energy).

The decision was made at the outset of this text to avoid advanced mathematics. Mathematics gives us a convenient shorthand notation with which to express Newton's second law or its fluid mechanics equivalent. For those who enjoy seeing the mathematics behind what follows are referred to Appendix A, where some details are given. Simpler examples that consist of the evaluation of formulae are to be found in Chapter 9. However, it is by no means essential to do mathematics to understand what follows; it is only necessary here to discuss what the terms in the balances represent and to use our knowledge of their dimensions followed by a knowledge of the magnitudes of the fundamental quantities from which the forces and accelerations are composed. The procedure in this section is thus to write down the balances (Newton's law is but one), and then identify various components of these balances and express these components in terms of fundamental quantities. We can then use dimensional analysis to gain an understanding of processes relevant to various phenomena in marine science. We need to list those quantities which can be regarded as fundamental to marine science. These are given in Table 2.1.

As we see from Table 2.1, acceleration has dimension LT^{-2} (metres per second squared). In marine science, particularly in meteorology, but also in most branches of oceanography, the rotation of the Earth plays an important part. The angular velocity of the Earth is about 7.29×10^{-5} rads per second and has dimension T^{-1}. Hence a velocity multiplied by the angular velocity of the Earth has dimensions $LT^{-1} \cdot T^{-1} = LT^{-2}$, the same dimensions as acceleration. Acceleration formed in this way is called Coriolis acceleration and plays a key role in most ocean dynamics. Perhaps the best illustration of Coriolis acceleration is motion on a steadily rotating turntable. Imagine drawing what you think is a straight line on such a turntable. (If you are old enough to possess a record player for playing vinyl discs, try this with a piece of chalk; if you have to borrow one, check with the owner first!) You will find that the line is curved. The 'fictitious force' that prevented the line from being straight and in accordance with Newton's first law is due to the Coriolis effect caused by the rotation of the turntable. You should notice that the line curves to the right for anticlockwise rotation, no matter what the direction of the drawn line. The Earth is always rotating, so that anything that moves on its surface will experience a similar force. Another way of generating acceleration is through (velocity)^2 divided by length. A quick check reveals this to be $U^2 L^{-1} = LT^{-2}$ (since velocity has dimensions LT^{-1}). Acceleration composed solely of velocity and length can be totally independent of time (if the velocity is steady, for example). A steady river flow is a good illustration of advective acceleration, but only if the river bends. As the water rounds a bend, each water particle is travelling with a constant speed, but there must be an overall

Table 2.1

Quantity	Unit	Dimensions
Mass	Kilogram (kg)	M
Length	Metre (m)	L
Time	Second (s)	T
Temperature	Kelvin (K)	Dimensionless
Velocity	Metres per second (m s^{-1})	LT^{-1}
Acceleration	Metres per second per second (m s^{-2})	LT^{-2}
Area	Square metre (m^2)	L^2
Volume	Cubic metre (m^3)	L^3
Discharge	Cubic metres per second (m^3 s^{-1})	L^3T^{-1}
Force	Newton (N)	MLT^{-2}
Pressure	Pascal (Pa)	$ML^{-1}T^{-2}$
Pressure gradient	Pascals per metre (Pa m^{-1})	$ML^{-2}T^{-2}$
Density	Kilograms per cubic metre (kg m^{-3})	ML^{-3}
Dynamic viscosity	Newton seconds per square metre (N s m^{-2})	$ML^{-1}T^{-1}$
Kinematic viscosity	Square metres per second (m^2 s^{-1})	L^2T^{-1}
Surface tension	Newtons per metre (N m^{-1})	MT^{-2}
Weight (same as force)	Newton (N)	MLT^{-2}
Angular velocity	Radians per second (s^{-1})	T^{-1}
Angular acceleration	Radians per second square (s^{-2})	T^{-2}
Vorticity	Radians per second (s^{-1})	T^{-1}
Circulation	Square metres per second (m^2 s^{-1})	L^2T^{-1}
Energy	Joule (J)	ML^2T^{-2}
Work (same as energy)	Joule (J)	ML^2T^{-2}
Power	Watt (W)	ML^2T^{-3}
Temperature gradient	Degrees per metre	L^{-1}

acceleration because two water particles at A (see Figure 2.1) will later be at point B, and although the flow is steady, these particles have changed orientation with respect to each other over the time it has taken for them to go from A to B. The same is true for all the fluid particles. This is perhaps the simplest illustration of the advective acceleration which is always present in a fluid (be it on a scale appropriate to the rotating Earth, or on a much smaller engineering scale where Coriolis effects are negligible). A shear flow as experienced by a current next to the coast or the sea bed is another example. Here, the two particles will drift apart as time progresses since the current closer to a boundary is slower. The flow is still steady and it is accelerating under the influence of advective acceleration. Time now for the definitions. *Advective acceleration* is acceleration due to the change in relative positions of fluid particles. *Point acceleration* is simply the rate of change of velocity with time, it is the acceleration experienced by you and I in an accelerating vehicle that is travelling in a straight line. It has dimension LT^{-2}.

To summarise, therefore, we have three separate kinds of acceleration; the total acceleration of an ocean current, tide or other oceanographic flow is the sum of these three effects:

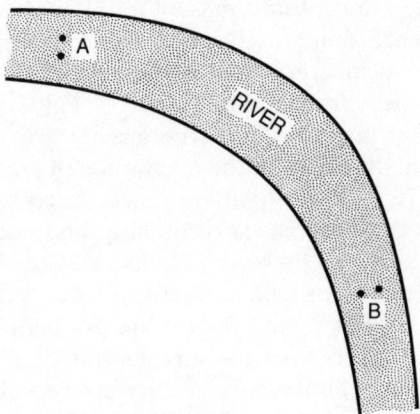

Figure 2.1 Adjacent particles flowing down a river.

acceleration = point acceleration
+ advective acceleration
+ Coriolis acceleration.

In terms of dimensions, the usual way to represent the three terms on the right-hand side is

$$(UT^{-1}) + (U^2 L^{-1}) + (\Omega U),$$

where U is an appropriate velocity magnitude, L is an appropriate length, T an appropriate time, and Ω is about 10^{-4} (the order of magnitude of the angular velocity of the Earth). To give a hint of the kinds of deductions that can be made from a comparison of dimensions, take a look at point acceleration UT^{-2} and Coriolis acceleration (ΩU). These terms will be about the same if $T^{-1} = \Omega$, that is, for time scales of about one day. This time scale covers tides; hence we deduce that both particle and Coriolis accelerations are important if we wish to model tides. For phenomena with time scales of months (e.g. Gulf Stream meanders), in general point acceleration can be ignored compared with Coriolis acceleration. On the other hand, for phenomena having time scales of seconds (e.g. surface water waves) Coriolis acceleration can be safely ignored compared with point acceleration. These of course are generalisations. Swell waves of exceptionally long period can be influenced by Coriolis effects; in other situations the physics can have complicating factors such as more than one characteristic velocity or length scale (as can happen near density interfaces), in which case there is perhaps a choice of scales and a choice of magnitudes of accelerations.

Let us now turn to the side of the equation that includes the forces. There are three, all very distinct, forces that act on a given volume of fluid. Think of this volume as in an arbitrary position, but of unit mass. First, the easiest to see is the force of gravity that we all experience. This gravity is not, in fact, true gravity but a combination of true gravity and centripetal force due to the spin of the Earth, which tends to throw our fluid volume away from the Earth's axis. This is called apparent gravity, but the good news is that apparent and true gravity are almost equal due to the centripetal force being one-thousandth the magnitude of gravity. So we have gravity, which for our purposes is a constant 9.81 m s^{-2} directed towards the centre of the Earth. The second force is due to the surrounding fluid exerting a pressure on our chosen volume. The force due to pressure is in fact due to the difference in pressure across the volume. The greater this difference, the greater is the net force due towards the lower pressure. We call this force the pressure gradient force. Finally, there is a force due to the frictional effects from the surrounding fluid. In the laboratory these frictional effects can be caused either by molecular action, called viscosity, or by eddy motion, called turbulence. Turbulence is still not very well understood. It is the irregular motion of a fluid which is such that there is continual mixing between it and the surrounding fluid, in contrast with laminar flow in which no such mixing occurs and the streamlines are well defined. Turbulence is most easily characterised by eddies that represent this mixing. (These eddies have the most important characteristic of transferring momentum at right angles to the mean flow, hence efficient mixing.) To say much more about turbulence would be inappropriate here as it can get very technical. Unfortunately, however, in the real ocean or atmosphere, virtually all frictional effects are caused by turbulence since turbulence is a much more violent effect than molecular friction (the factor is at least ten thousand and often greater).

<u>These then are the three forces: a force due to gravity that always can be assumed to act towards the centre of the Earth, a force due to pressure gradient, and a force due to turbulence.</u> These latter two act in a direction that varies with space and time, and are dependent on factors such as the speed and direction of currents.

The dimensions of the 'force per unit mass' are LT^{-2}. Gravity, usually denoted by g, is an acceleration which is automatically LT^{-2}. The pressure gradient force is pressure gradient divided by density $= ML^{-2}T^{-2} \div ML^{-3} = LT^{-2}$, as required. Finally, the turbulence term is a little more tricky. In fact, in its simplest form we characterise turbulence via a constant eddy viscosity which is an enhanced kinematic viscosity. The turbulent stress is this eddy viscosity multiplied by density times a velocity gradient. Hence turbulent stress has dimensions $L^2T^{-1} \cdot ML^{-3} \cdot LT^{-1} \cdot L^{-1} = ML^{-1}T^{-2}$. The force per unit mass due to turbulent stress is now the gradient of this turbulent stress divided by density. This leads to dimensions $L^{-1} \cdot ML^{-1}T^{-2} \div ML^{-3}$, i.e. LT^{-2}, as required.

This last group of terms, the force due to turbulence, is worth further consideration. Suppose we denote by v the magnitude of the eddy viscosity (the fact that this is not a directly measurable quantity used to be a bone of contention; it is less so now because it is more widely recognised as an idealisation useful for dimensional analysis rather than a real quantity). In terms of v, and a length scale L the force per unit mass due to turbulence is $vL^{-2} \cdot LT^{-1}$ (recall that v has dimension L^2T^{-1}; the last grouping, LT^{-1}, is interpreted as a typical velocity). In marine science, there is a difference in magnitude between horizontal and vertical quantities. This is seen by comparing, say, the depth of the ocean to its width (a factor of one thousand is involved). Sometimes this is less relevant, for example, when dealing with an estuary; however, the ratio of depth to horizontal length is called the aspect ratio, a term first coined in aeronautical engineering. The letter D is usually used for a typical vertical length (D for depth), whereas L means horizontal length. The force term due to turbulent stress can thus be written vUL^{-2} or vUD^{-2}, depending on whether we wish to address horizontal or vertical phenomena.

We are now in a position to write down the force balance (Newton's second law) in dimensional terms. A comparison of various pairs then leads to *dimensionless* groups of fundamental quantities

$$\text{force per unit mass} = \text{acceleration}.$$

In words, this is

$$\text{pressure force} + \text{gravity} + \text{turbulence} = \text{Coriolis acceleration}$$
$$+ \text{advective acceleration}$$
$$+ \text{point acceleration}.$$

We have already given a broad hint that Coriolis acceleration is particularly important. Let us first examine the consequences of a balance between the Coriolis term and turbulence. In order for Coriolis acceleration (ΩU) to be compared with turbulence vUL^{-2} (or vUD^{-2}), we examine the magnitude of

$$\frac{vUL^{-2}}{\Omega U} \quad \text{or} \quad \frac{vUD^{-2}}{\Omega U},$$

i.e.

$$\frac{v}{\Omega L^2} \quad \text{or} \quad \frac{v}{\Omega D^2}.$$

The dimensionless grouping $v/\Omega L^2$ (or $v/\Omega D^2$) is called an Ekman number. The version containing L is the horizontal Ekman number; that containing D the vertical Ekman number. We will have much more to say about Ekman numbers in Chapter 6. Another important balance occurs when the Coriolis acceleration (ΩU) becomes of similar magnitude to the advective acceleration (U^2/L). In order to compare *these* two terms, we need to examine the magnitude of the ratio

$$\frac{U^2/L}{\Omega U} = \frac{U}{\Omega L}.$$

This dimensionless number, $U/\Omega L$, is called the Rossby number. Again we shall return to this later when discussing model details. Note here, however, that the angular speed of the Earth, Ω, is usually replaced by $f = 2\Omega \sin(\text{latitude})$, which is twice the vertical component of the Earth's angular velocity, in the definition of Rossby number. This is because it is f that plays a major role in modelling. The Rossby number is then $U/(fL)$.

2.3 Other balances

Another obvious physical law concerns the destruction or creation of matter. It does not happen in marine science; matter is neither created nor destroyed. This is called the equation of continuity in fluid mechanics and leads us to deduce that a typical vertical velocity is UDL^{-1}, where U is a typical horizontal velocity, D is a typical depth and L a typical horizontal length. To convince yourself of this, imagine a box whose horizontal dimensions are L and whose vertical dimensions are D. Then the net flow in the sides will have magnitude UD, whereas flow out of the top or the bottom will have magnitude WL (where W is a typical vertical velocity). Since WL and UD must be of the same order (by conservation of mass), then W must be UDL^{-1}. The ratio D/L is called the *aspect ratio* (a similar name is used in aeronautical engineering) and is commonly about 10^{-3} in the oceans. We therefore deduce that vertical velocities are about one-thousandth of their horizontal counterparts. This is not to say that we can always ignore vertical currents. Upwelling and, to a lesser extent, downwelling have an important role in biological cycling and production processes. We shall return to this later.

The last equations we shall consider are those that express the conservation of temperature and the conservation of salinity. In the oceans, the temperature and salinity are related to the density via a relationship called the equation of state. The atmosphere is of course a gas and not only is there is no salinity, but the equation of state is of an entirely different character due to the difference in sizes of the constituent molecules. The simplest modelling assumption would be to assume the validity of the equation of state for a perfect gas, corrected for the imperfections of moist air. In the sea the conservation of temperature (or salinity) consists of equating a rate of change of temperature (or salinity) to its dissipation via diffusion. This diffusion plays an analogous role to that of eddies in fluid turbulence. Instead of v, the eddy viscosity, we have eddy diffusivities of heat (for temperature) and salt (for salinity). In turbulence, a comparison between advective acceleration and turbulent stress force leads to the dimensionless quantity UL/v, the turbulent Reynolds number, or its inverse v/UL, the Taylor number. These are not as important in marine science as they are in laboratory fluids because of the dominant presence of Coriolis acceleration. Nevertheless, their temperature and salinity counterparts (the temperature counterpart of the Taylor number is often called the Prandtl number; I know of no universally accepted name for the dimensionless ratio that involves salinity) are important when comparing the relative effects of diffusion and other non-diffusive processes. Diffusion is so important it has a chapter to itself.

2.4 Internal balances

The previous sections have addressed balances in the ocean that imply that all phenomena are basin wide, estuary wide or even ocean wide. Characteristic length, time and velocity scales are taken to be uniform over an oceanic region of interest. There are many phenomena that are restricted to particular zones. In the ocean, most of these zones are bounded by density interfaces of one kind or another. Oceanographers have been aware of the pycnocline (density interface), thermocline (temperature interface) and halocline (salt–water interface), where these interfaces are primarily horizontal, dividing two bodies of water that ride over each other. <u>Vertical interfaces are called fronts,</u> but obviously pycnoclines can become fronts and vice versa. Near a pycnocline (the word actually covers thermoclines and haloclines too) one might expect the horizontal velocity to change rapidly with depth. The density also changes. Given that density changes with depth, its gradient, divided by an average ambient value gives a measure of this change. The quantity

$$\frac{\text{gradient of density}}{\text{density}}$$

has dimension L^{-1} (the density part cancels). In many circumstances, the ocean can be assumed to consist of two layers, the top layer representing the mixed layer, and the deep layer, with the interface between the layers being the thermocline. In this case, the above quantity takes the equivalent form

$$\frac{\text{difference in density}}{\text{density}}$$

(see Chapter 5). If this quantity is multiplied by gravity (dimension LT^{-2}, acceleration) we obtain a quantity of dimension T^{-2}, that is, the square of the frequency. This frequency is proportional to the square root of the density gradient (or difference). It is representative of the degree of stability of the ocean (that is, the likelihood or tendency for vertically displaced parcels of water to return). This stability frequency, given the symbol N, is called the Brunt–Väisälä frequency. N is a frequency (dimension T^{-1}). The vertical gradient of horizontal velocity (that is, how horizontal currents change with depth) also has this dimension since it is velocity LT^{-1} divided by L. Note that we do not distinguish between vertical and horizontal length scales here since we are discussing relatively small-scale phenomena in a boundless ocean. If we take the ratio of N to this vertical gradient in horizontal current (vertical shear) we have a dimensionless quantity that compares how stable a column of water is with the magnitude of the vertical shear of the (horizontal) currents. Because square roots are involved in defining N, the Brunt–Väisälä frequency, this dimensionless ratio is normally squared. It is given the name Richardson number, R_i (or, more correctly, the gradient Richardson number, since there is a slightly different but more fundamental one that can be defined in terms of turbulence itself rather than gradients in the mean current). We shall discuss the Richardson number when we discuss boundaries in Chapter 4. It is worth noting here that if R_i is less than zero, density variations enhance turbulence; if, on the other hand, R_i is greater than zero, density variations reduce it.

2.5 Other dimensional quantities

It is often constructive to consider generalisations that involve quantities that in fact possess dimensions but in other respects are as general as the non-dimensional ratios defined earlier. As an illustration, let us go back to the boundless ocean. If a large body floats in such a boundless sea, and is given a speed U, then, as it begins to move, the Coriolis acceleration acts to move it to the right of its direction of initial motion (in the northern hemisphere; and left in the southern hemisphere; we steer well clear of the equator). If the effects of friction are ignored, then the body will move in a large circle. The radius of this circle will be U/f, where f is the local value of the Coriolis parameter ($2\Omega \sin(\text{latitude})$), where Ω is the angular speed of the Earth. This length scale is fundamental. It is the characteristic length associated with horizontal motion in a homogeneous sea and is given the name the Rossby radius of deformation, or simply the Rossby radius. The circle described by our fictitious floating body is called an inertial circle, and the Rossby radius is the radius of this circle. Now, let us get a little closer to reality and consider a large ocean current, for example, our old friend the Gulf Stream. Suppose this has a characteristic length scale L which is determined by its driving mechanisms (e.g. wind stress). If the Rossby radius is compared to this characteristic Gulf Stream length we obtain a dimensionless quantity $(U/f)/L$, which is $U/(fL)$, another old friend, the Rossby number. This provides an alternative derivation of the Rossby number that is less dependent on an understanding of advective acceleration. Another phenomenon yet to be mentioned here is waves. If a water wave travels in a shallow regime of depth h then the speed of travel of the wave is $\sqrt{gh}$, where g is, as usual, acceleration due to gravity. This is the speed of travel of tides as they propagate over the continental shelf. If you can be convinced that the speed of travel of such waves can only depend on the local depth and the restoring force for the water surface (gravity) then this formula follows straight away from dimensional analysis

$$\text{speed} = g^A h^B = (LT^{-2})^A L^B,$$

from which, since speed has dimension LT^{-1}, we obtain simultaneous equations for A and B as follows:

$$1 = A + B,$$
$$-1 = -2A,$$

from which $A = \frac{1}{2}$ and $B = \frac{1}{2}$, as required. If there is a change of density with depth, then the picture is more complex. Some headway can be made quickly by assuming a two-layer sea, in which case g is replaced by reduced gravity g' which is defined by

$$g' = \frac{\rho_1 - \rho_2}{\rho_1} g,$$

where ρ_1 is the density of the upper layer and ρ_2 the density of the lower layer, and instead of h the depth of the upper layer is used. The wave speed calculated from these quantities is $\sqrt{g'H}$, where H is the depth of the upper layer. Since g' is generally about $g/(500)$ and H is less than half of h, this stratified long wave speed is much less than the homogeneous long wave speed. Typical values are 20 m s^{-1} for the homogeneous long wave speed and 0.5 m s^{-1} for the stratified version. These latter much slower speeds are important in the dynamics of internal tides which are thought to be particularly significant in areas such as the Mediterranean, the Caribbean and around Australia. Let us return to the body floating on a large boundless sea. The approriate speed may well be the long wave speed usually labelled c (we have written $c = \sqrt{gh}$). The Rossby radius of deformation is therefore c/f. If we now turn our attention to a body floating on the thermocline, then the Rossby radius as just defined will not be appropriate. Instead, we need a quantity based on c_1, where $c_1^2 = g'H$. That is, the characteristic speed is the much slower internal wave speed. The characteristic length is thus c_1/f, which is much smaller and is called the internal Rossby radius. It is through this mechanism that the Coriolis effect can be felt even in an estuarial environment (see Chapter 6 and the publications of Dyke, 1980, 1987.

Waves in water are responsible for the generation of associated currents, although these wave-induced currents are periodic, or very nearly so. In the sea, superimposed on a wave regime, there is often a current that is generated from another completely separate source, perhaps a river outflow or a wind-induced flow. If the characteristic speed of this current is U, then a dimensionless number can be generated by comparing U with the wave speed c. Because square roots are involved the ratio is usually squared and we arrive at the quantity U^2/gh, which is called the Froude number. This number is very important in hydraulics, in particular for classifying surface phenomena such as bores and hydraulic jumps. In a stratified two-layer estuary or river, a comparison between U and the internal wave speed is made, whence the dimensionless quantity is $U^2/g'H$, an internal or densimetric Froude number. This latter Froude number is important to consider when discussing billows on internal waves and laboratory simulations of them. These play a role in mixing across the thermocline.

2.6 Examples of dimensional analysis

In order to see how dimensional analysis actually helps us to set up models, let us do two examples. By necessity, some material in this section is of a slightly more technical nature. I do not think that any apology for this is necessary.

2.6.1 The Gulf Stream

The first stage in the process is to designate appropriate values to the quantities U, L and v. For the Gulf Stream, L is either taken as its width, which is typically

100 km, or 10^5 m, or perhaps even its length, which is ten times this (10^6 m). Along the Gulf Stream itself, the velocity is typically 1 m s^{-1}. Across the Gulf Stream, the typical velocity is one-tenth of this (0.1 m s^{-1}). There is no controversy about the Coriolis parameter – this is 10^{-4} s^{-1}.

There is much more controversy, however, about the value of v, which for horizontal turbulence is usually taken to be of magnitude 10^6 m^2 s^{-1}, but for vertical turbulence, 10^{-1} m^2 s^{-1} is more usual. The next stage is to calculate the magnitudes of the appropriate dimensionless numbers. For this particular problem, the horizontal Ekman number and the Rossby number are both calculated. For motion along the Gulf Stream with the given values of the parameters, both the horizontal Ekman number v/fL^2 and the Rossby number U/fL are 10^{-2}. The appropriate conclusion is therefore that the effects of friction (Ekman number) and the effects of advection (Rossby number) are not important to the motion of the stream in the direction along the stream. On the other hand, for motion across the stream, although the Rossby number remains small at 10^{-2}, the horizontal Ekman number becomes close to unity. Frictional effects are thus important to cross-stream dynamics. However, one needs to be mindful of what assumptions have been made. For example, if the Gulf Stream bends or meanders, then the magnitude of L is open to question. If U remains large at around 1 m s^{-1}, and L is 100 km, then the Rossby number becomes large. It is therefore likely that advective terms (sometimes called inertial effects) are important in order to resolve the dynamics of Gulf Stream meanders. Detailed mathematical modelling has indeed revealed this to be true. As the water of the Gulf Stream travels along the coast of the USA, its dynamics become more and more complex as the controlling mechanism transfers from being friction to being advection. In order to resolve the details, it is not good enough merely to do dimensional analysis. Instead, the equations must be solved. This is usually done by numerical means, which is the subject of the next chapter. Here is another example showing a more arithmetical approach to dimensional analysis.

2.6.2 *Parametrising bottom friction*

Friction, or drag, caused by the action of the sea bed on the water just above it is a force. As such, therefore, it has dimensions MLT^{-2}. Physically, one expects this force to depend on the speed of the water, $U = LT^{-1}$. It must also vary with the typical length scale L; the longer L is, the larger is F. The more viscous (or turbulent) the water is, again the larger is F. Finally, the larger the fluid density, ρ, the larger is F. Now we use what is commonly called the Buckingham pi theorem. This takes the form of equating F to powers of U, L, ρ and μ (the dynamic eddy viscosity). That is, we write

$$F = L^A U^B \rho^C \mu^D.$$

Now, from Table 2.1, the dimensions of ρ are ML^{-3}, and the dimensions of μ are $ML^{-1}T^{-1}$. From these relationships, writing $F = MLT^{-2}$ and $U = LT^{-1}$, we obtain

$$MLT^{-2} = L^A(LT^{-1})^B(ML^{-3})^C(ML^{-1}T^{-1})^D.$$

Equating powers of M, L and T to each other gives the three equations:

M: $\quad 1 = C + D,$
L: $\quad 1 = A + B - 3C - D,$
T: $\quad -2 = -B - D.$

These cannot be solved (three equations in four unknowns). But writing everything in terms of D, we get

$$C = 1 - D, \quad B = 2 - D, \quad A = 2 - D,$$

so that

$$F = \rho L^2 U^2 f\left(\frac{\mu}{\rho L U}\right),$$

where the last term on the right is an arbitrary function (the combination $\mu/\rho L U$ is dimensionless). Now it turns out that there are good reasons for letting $L^2 f(\mu/\rho L U)$ be a constant for a given regime. This constant is usually called C_D. Hence we arrive at

$$F = \rho C_D U^2,$$

which is the usual quadratic friction law. If the current is a vector, i.e. it has a direction associated with it, then the current and drag are in opposite directions and we write

$$F = -\rho C_D \boldsymbol{u} |\boldsymbol{u}|.$$

This is the usual form of the drag law.

Chapter Three

Numerical methods

3.1   Introduction

It may seem rather bizarre to include a chapter on numerical methods in a book which, from the outset, has guaranteed very little mathematical content. However, the main purpose of this chapter is to *discuss* the various numerical methods and use of software without going into too much mathematical detail. Some practice in actual mathematical modelling is postponed until Chapter 9, and even there it is very gently paced. Even after this, it is recognised that the reader will in no way be able to build his or her own models. However a basic understanding will, it is hoped, be gained.

The equations that describe the behaviour of the ocean are *differential* equations; that is, they involve rates of change of the basic quantities such as velocity, temperature, pressure, etc. Computers cannot deal with continuous quantities as such. By their very nature they deal with discrete information, the famous 0, 1 (off, on). This was not a crucial problem in the early 1960s and before when models were simple enough to be solved analytically by a variety of exact mathematical means. Now, however, modelling is no longer the sole province of the mathematically inclined; the focus is on the public understanding of models. How then are equations involving continuous quantities reproduced and solved discretely?

3.2   Numerical methods in oceanography

There are two basic types of numerical model. One is based on finite differences; the other on finite elements. Since we are not focusing on detailed mathematics it is not necessary to go into fine detail. Instead, enough will be said for the reader to glean the essential features of both types.

Finite differences are based on substituting differences for derivatives. To understand this, look at Figure 3.1. This depicts a graph of a function which might be one component of velocity or surface elevation. It is shown to vary continuously. Suppose we wish to know the gradient of $f(x)$ at a particular point (labelled x_0 in Figure 3.1), this would be the derivative of $f(x)$. It is the slope of the tangent at x_0. Suppose we

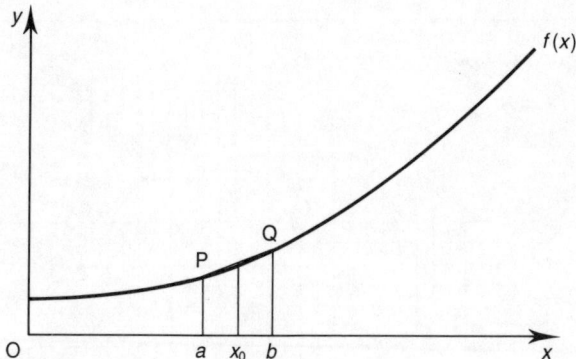

Figure 3.1 The graph of a function $f(x)$, with chord PQ.

have measurements of $f(x)$, or equivalently values from another source, at $x = a$ and $x = b$ but no information about $f(x)$ at x_0 itself. We can approximate to the slope at x_0 by finding the slope of the chord PQ which can be calculated straightforwardly from knowledge of the values of $f(x)$ at a and b, as well as the differences between b and a, i.e. $(b - a)$. This is the underlying principle behind finite differences. In fact, there are three types of difference: a forward difference where values at x_0 are based on information to the *left* of x_0 (i.e. $x \leq x_0$); a backward difference where values are based on information to the *right* of x_0 (i.e. $x \geq x_0$); and finally, a centred difference where values are derived from information on both sides of x_0. Backward differences are not widely used, but forward differences are particularly useful when dealing in the time domain where the need is to predict at a later time from that which is known at an earlier time. However, centred differences provide the most accurate approximations, even though the mathematics is more complex. In a mathematical model all gradients, in x, y, z as well as in time t, are replaced by these differences. Practically, the estuary, sea or ocean that is of concern and that one wishes to model is overlaid with a square grid (it can be other shapes but this is unusual). The intersections of the grid called the mesh points are the two-dimensional versions of the a and b of Figure 3.1. Figure 3.2 shows a grid over the European continental shelf. In the most straightforward finite difference scheme, values of velocity and surface elevation will be given at the grid points. Also shown in Figure 3.2 are crosses that indicate those locations for which inputs from the meteorological model operated by the UK Meteorological Office are available. Since these values are vital for the model as boundary conditions (see Chapter 4) but do not coincide with the grid intersections themselves, some interpolation is necessary. The nature of the equations is such that they are predictive; which means mathematically that, in the time domain, a forward difference is appropriate. This means that from values at a given time over all grid points it is possible to step forward in time and predict the values of velocity and surface elevation at the next time step and thence at subsequent times. In a real

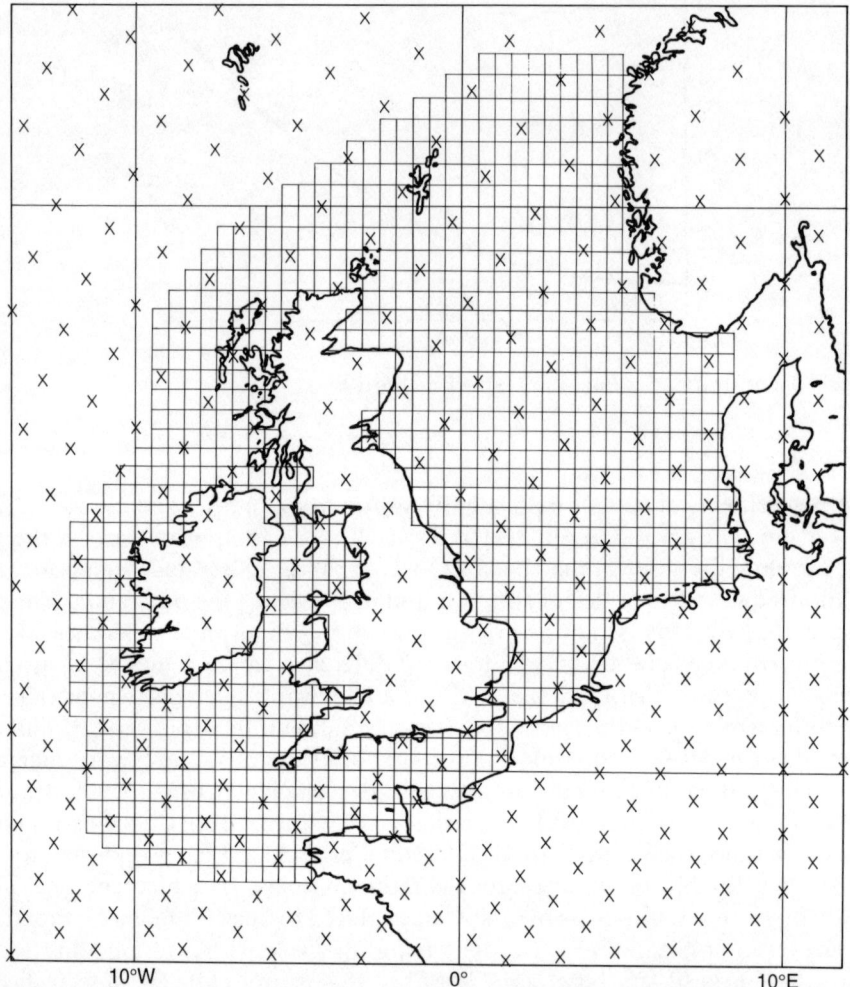

Figure 3.2 Finite difference mesh of the continental shelf model with grid points (×) of the 10-level model of the atmosphere. From Flather (1979). Reproduced with permission.

numerical simulation, this prediction does not occur at once at all grid points. The nature of the prediction depends on the variety of difference scheme adopted.

Let us now consider the third dimension. In oceanography (and meteorology) the third coordinate is vertical and so, because the aspect ratio (the ratio of typical vertical to typical horizontal length scales; see Chapter 2) is small, the size of any steps along this axis is much smaller than any horizontal counterpart. For this reason, the purely horizontal two-dimensional model is useful for many practical purposes. However, the capacity and speed of modern computers means that it is now possible to contemplate a three-dimensional grid. Such models are still thought of as 'layered', since the vertical structure of the ocean has particular features (the thermocline, the mixed layer and the bottom boundary layer, for example) which lend themselves naturally to a layer concept. There is a useful alternative to using layers. The vertical structure of a typical sea is quite different at different depths. For example, near the bed and close to the thermocline there is usually a lot of fine structure that would need very small vertical step lengths to resolve. Using vertical modes would enable this resolution to be locally fine in an efficient manner. Modal methods depend on using continuous functions to represent how the variables vary through the depth. In this context, a mode can be defined as a particular (fixed) variation of a variable with the vertical. For example, the first (usually called the zeroth by mathematicians) mode is no variation at all, the second represents a two-layer situation, etc. It turns out that only a few functions are required in order that a variation of velocity with depth (say) can be successfully mimicked. To those that understand mathematics, these vertical modes sum in the same way as a Fourier series. It is rather like a picture that contains a bewildering variety of beautiful colours rather startlingly being made from combining merely the three primary colours (red, green and blue) but using varying intensities. The principal advantage of using modes is their ability to allow variables such as velocity to depend continuously on the depth coordinate. This enables precise values to be given at *any* vertical location. This ability for continuous dependence is also a characteristic of finite elements.

The principle behind finite elements is to cover a region with small faces (usually triangular rather than square or rectangular). These faces are small enough so that the variables, velocity, pressure and the like, vary only very little over a single face. This enables key variables to be expressed by simple linear (straight line) expressions. The overall balances are satisfied by summing all the (approximated) variables over all faces. Pure finite element procedures cannot be applied to any but the simplest problems. However, there is a very useful hybrid technique that uses concepts from both differences and elements called the Galerkin weighted residual method. The details of this method are beyond the scope of this text and the interested reader is referred to Davies (1987) and the source text of Finlayson (1972). The weighted residual method is sometimes referred to as a 'spectral method' because of the treatment of variations in the vertical that involves adding together lots of simple solutions (modes) to generate a complex one. Meteorological modellers often use spectral techniques because of the closed spherical nature of their problem, there are no coasts in the atmosphere. However, fortunately for the non-mathematically inclined, there is no need to dwell on details since many of the features of models that are considered important to highlight here are not dependent on them.

3.3   Errors

An error in a numerical simulation is any reason why it does not work. Of course, a model may not work because the parameters of the model have been given the wrong values, but we are not considering this here. Nor are we considering human errors such as mistyped instructions or misplaced minus signs, we can forget about all these. No, here we are talking about errors induced by the numerical procedure itself. One type of error which is much less common than it used to be is round-off error. This kind of error occurs because only a certain accuracy can be guaranteed in the arithmetic the computer carries out. For example, the constant π (3.141 59...) can perhaps only be held to a certain number of decimal places. Although software must be designed to minimise round-off error, and it is a much less serious error than it used to be, it is still very important to carry out the arithmetic of a numerical method in an order that enables this minimisation to take place.

A potentially serious error is one called truncation error. This kind of error is caused by the numerical procedure itself, and is due to the actual problem being solved by the computer not being the real problem. The slope of the chord PQ in Figure 3.1 is not precisely the gradient of $f(x)$ at x_0. In order to reduce truncation error, the approximations have to be very accurate. In order to be very accurate, the numerical schemes usually have to be complicated, and the more complicated the scheme the larger the computer needed to run it and (usually) the longer it takes to run. There are several trade-offs in this hierarchy. At one end of the scale, we have 'quick and dirty' models which at least produce an answer, but it can be inaccurate. Beware the 'quick and dirty' model with a glamorous front end, promising much with its colour graphics and user-friendly menu, the modelling equivalent of the Greek sirens! At the other end of the scale there are very well written sophisticated models that require supercomputing power to run. At this end of the market there is an additional trade-off; there is a choice between obtaining either quick results less accurately or accurate results more slowly. The chess playing reader is probably familiar with the highest-level programs that take hours or more per move, whilst lower levels produce less sophisticated games more quickly. It is a similar trade-off.

The catalogue of errors does not end with round-off error and truncation error. When dealing with predictive equations, another problem is stability. This, technically, is not really an error since it is an inherent property of the scheme chosen. Calling it an error is rather like blaming someone for his or her height. Instability is the tendency for the solution to a model to oscillate rather than to settle down to an acceptable value. If this oscillation gets larger and larger as time progresses, then there is nothing that can be done apart from completely redrafting the problem using a different (*more stable*) finite difference scheme. More normal, however, is an oscillation that remains finite. A way to deal with this is to increase the amount of averaging in space so that the oscillation is averaged out as the calculations of the scheme progress. This can work quite well, but increasing the averaging is another way of saying increasing the truncation error, and to the lay user, this extra truncation error is very difficult to detect; it is a modelling sleight of hand to introduce extra error in order to dampen out instability. This is a really dirty trick since it introduces one inaccuracy in order to mask poor performance. It is a car dealer painting over the rust, or pouring gunk into the sump to 'cure' the oil leak. To detect such underhand trickery in modelling is not easy even for the professional modeller. Fortunately it is

Numerical methods 23

rare to encounter blatant cheating in models, but perhaps there should be some 'trading standards officers' equivalent to those in the motor trade (in some branches of engineering, this is indeed beginning to happen).

It is an unfortunate fact that the truncation error induced by the numerical treatment of the equations in finite difference form has precisely the same mathematical form as the friction. It is therefore normally impossible to distinguish real friction (fluid turbulence) from its numerically induced clone. The controversy over the actual values of eddy viscosity used in numerical models has been a focus of many interesting discussions at modelling research meetings. For example, modellers of shallow seas, whose grids are finer and the truncation error smaller, are scathing in their criticism of deep-sea ocean wide modellers who use large grid sizes and consequently induce larger truncation error into their schemes. The shallow sea modellers refer to the ocean in the deep-sea models as 'treacle' as a reference to the enhanced numerical error (truncation error) being indistinguishable from the very high (eddy) viscosity. They are subsequently often mistrustful of the results from deep-sea models given the remarks about the ability of large truncation error to mask instability. Having said this, newer deep-sea models have a fine enough grid size to refute these criticisms (see Chapter 5). In this chapter, the examples will demand some knowledge of mathematics. If this is too daunting, the following section may be safely omitted without affecting what can be gained from the rest of the book.

3.4 Examples using numerical methods

The following examples are incomplete since boundary conditions, crucial to the implementation of all models, have yet to be covered.

3.4.1 *A semi-implicit finite difference scheme*

This example is loosely based on a paper by Backhaus (1983). Let us focus on the horizontal discretisation in finite difference form. The grid shown in Figure 3.3 is the fully staggered grid, sometimes known as the Arakawa C grid. It will be noticed that there are three kinds of points, denoted by plus symbols, cross symbols and blob symbols. Surface elevation is evaluated at plus points. Since east–west velocity is computed from east–west differences (gradients) in elevation, it is convenient for the cross symbols (for east–west velocity) and the plus symbols (elevation) to alternate along the grid lines that travel east–west. Similarly, the blob symbols (for north–south velocity) and the plus symbols alternate along the grid lines that run north–south. This staggered grid is therefore entirely natural.

The differential equations obeyed by a sea contain either first-order or second-order derivatives. In order to understand the operation of the scheme, we need to see how these derivatives are approximated by finite differences. As shown in Figure 3.3, the spacing of the points in the east–west direction is Δx, and the spacing of the points

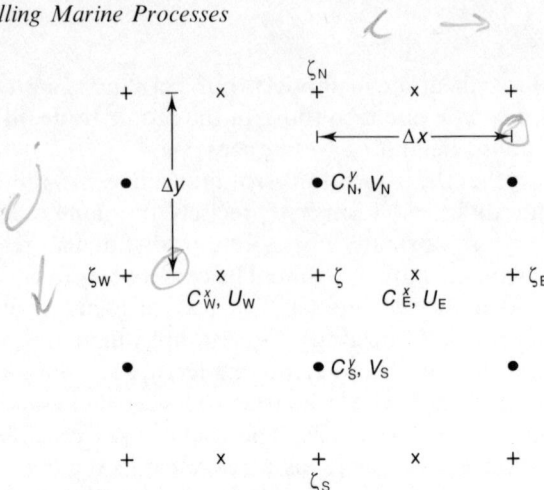

Figure 3.3 Richardson lattice: spatial arrangement of coefficients and variables. + pressure (ζ); × U-component; ● V-component.

in the north–south direction is Δy. This particular model contains no second-order derivatives, and so we take as our example the discretisation of the equation that governs mass continuity (the conservation of mass or 'nothing can be created or destroyed in an arbitrary volume of the sea'). This equation takes the form

$$\frac{\partial \zeta}{\partial t} + h \frac{\partial U}{\partial x} + h \frac{\partial V}{\partial y} = 0.$$

Let us denote the value of U at the point where $x = i\Delta x$, $y = j\Delta y$ by $U^s_{i,j}$ (the superscript s will be explained later), then the derivative of U with respect to x (that is, the rate of change of U as x varies) can be approximated by

$$\frac{\partial U}{\partial x} = \frac{U^s_{i+1,j} - U^s_{i-1,j}}{2\Delta x}.$$

In a similar fashion, if the value of V at the point where $x = i\Delta x$, $y = j\Delta y$ is denoted by $V^s_{i,j}$, then the derivative of V with respect to y (that is, the rate of change of V as y varies) can be approximated by

$$\frac{\partial V}{\partial y} = \frac{V^s_{i,j+1} - V^s_{i,j-1}}{2\Delta y}.$$

We denote by $h_{i,j}$ the value of the depth of the sea, h, at the location $(i\Delta x, j\Delta y)$. The approximation used for the first term, $\partial \zeta/\partial t$, is slightly different because time is involved. The whole basis of using finite differences as an approximation to the equations is to be able to predict what will happen to the variables u, v and ζ given what has and is happening to them. This means, as far as the mathematics is concerned, that we need to use forward differences whenever possible to discretise the time rates of change. Thus we replace $\partial \zeta/\partial t$ by

$$\frac{\partial \zeta}{\partial t} = \frac{\zeta^{s+1} - \zeta^s}{\Delta t},$$

where the notation $\zeta(t) = \zeta(s\Delta t) = \zeta^s$ and $\zeta(t + \Delta t) = \zeta((s+1)\Delta t) = \zeta^{s+1}$ has been adopted. This now explains the presence of s in U^s and V^s. It expresses the fact that U and V are evaluated at the time $s\Delta t$. The original equation for mass conservation is thus replaced by a difference equation of the form

$$\zeta^{s+1} = \zeta^s - \frac{\Delta t}{2\Delta x} h_{i,j}(U^s_{i+1,j} - U^s_{i-1,j}) - \frac{\Delta t}{2\Delta y} h_{i,j}(V^s_{i,j+1} - V^s_{i,j-1}),$$

where, as can be seen by the superscript s on all variables on the right-hand side of the equation, all these terms are evaluated at the earlier time step. This equation is (approximately) valid at each interior point of the domain, as are similar discretisations of the momentum equations. Therefore one can, at least in principle, solve them. In Chapter 9 examples are provided in which numerical values are substituted for all these symbols and for those uncomfortable with theory, life is breathed into the equations. Before any solving can take place, however, one has to deal with boundaries such as coastlines and the sea bed, and with the open boundaries at the edge of the chosen domain. These are the subjects of the next chapter.

3.4.2 A finite element method for tides in the western Atlantic

The details, such as they are, of this model are taken from several papers by the US researchers Johannes Westerink and Rick Luettich, notably Westerink et al. (1994). That paper contains a graded triangular mesh of staggering complexity. It is reproduced as Figure 3.4.

The model is two-dimensional, but uses spherical coordinates that cling precisely to the curved surface of the Earth. The natural coordinates chosen are longitude and latitude, and although these coordinates, being curvilinear, give rise to extra unfamiliar terms in the governing equations, conceptually the idea is simple. The two-dimensional slab of ocean is divided into triangles. All variables are assumed to be independent of depth, so that all that needs concern us is how to approximate those variables associated with tides over each triangle (the question of boundary conditions is left until Chapter 4). The triangles are smallest in those parts of the region where the variations are largest, and conversely they are largest where there is very little variation. There is a reason for this! In regions of high variability, there is a lot of structure in terms of changing currents and sea surface elevation. In order to capture this variation, many small triangles are required. On the other hand, in regions where variables are more or less constant, one large triangle that in effect connects the more interesting regions together will suffice. For tides, it is suspected that the shallower regions near the coast and close to the islands are the regions of high variability. Within each triangle, therefore, variables such as the horizontal tidal velocities and the tidal elevation (with respect to the geoid; that is, the equivalent to the standard tidal

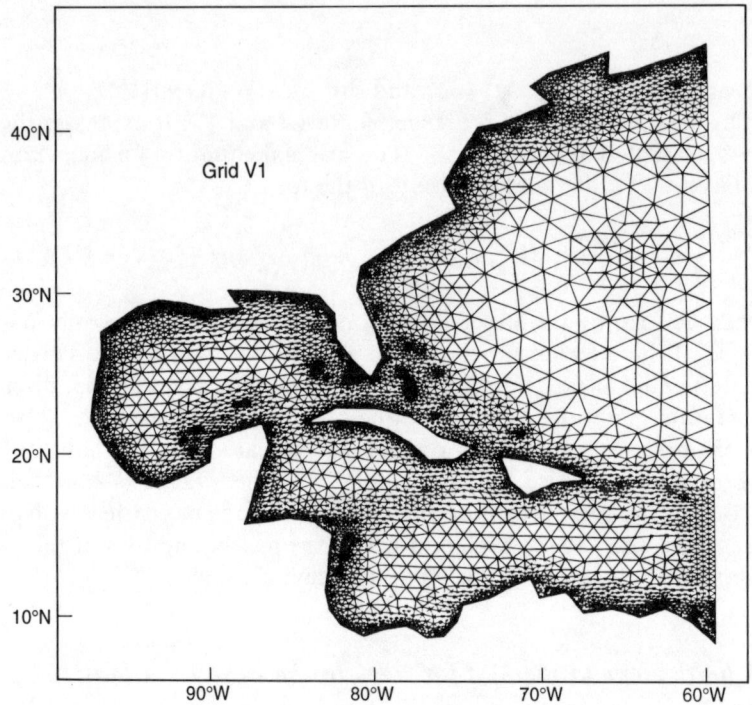

Figure 3.4 An unstructured finite element grid. From Westerink *et al.* (1994). Reproduced with permission.

elevation except that a spherical coordinate system is being used) can take a particularly simple form. Westerink *et al.* use simple trigonometric functions; in papers that utilise Cartesian coordinates, linear functions are popular. The important criterion is to ensure that all variables remain continuous at the borders of each triangle. For Westerink and his colleagues, this meant the elevation and velocities associated with the tide. This is consistent with the preoccupation of the paper, which was an accurate prediction of the tidal heights in sensitive areas such as around the land masses, particularly the islands.

If it is required to predict higher-order quantities such as tidally induced residual flow, then it essential for the expressions for the variables in each triangle to be sophisticated enough for these quantities themselves to be continuous at the edges of the triangles. This usually means that the functions that represent each variable in any particular triangle need to be more complicated. It is sometimes the case that shapes other than triangles are used in finite element simulations. The rectangle is a quite popular choice, when comparison with finite differences can be made easier, and near the coast, coordinates that fit the coastline more closely (termed isoparametric coordinates) can be very accurate and worth the extra effort. It remains the case that

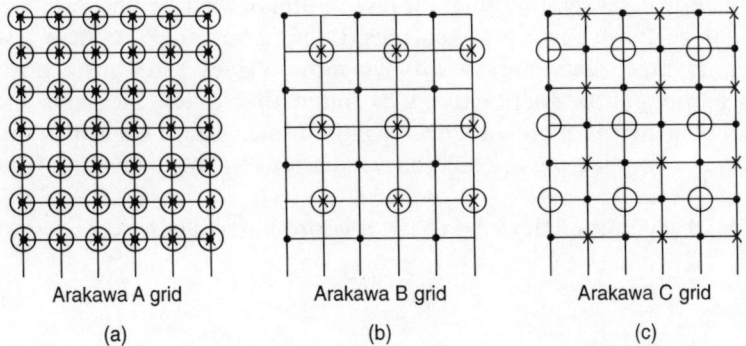

Figure 3.5 The three Arakawa grids. · elevation (ζ), ○ northwards current (v), × eastwards current (u).

finite differences are more popular than finite elements, mainly because of the bookkeeping associated with keeping track of the numbering of elements in regions where the shape is complex, but also because non-linear terms in finite element schemes lead to unacceptably high demands on computing power compared with finite difference schemes of equivalent complexity.

3.4.3 Finite difference grids for ocean models

No apology is necessary for returning to finite difference methods, for these dominate present-day numerical modelling. The history of the numerical modelling of ocean currents is not long – it dates from the mid-1960s – but several types of finite difference grid are in use and we take this opportunity to introduce the main varieties. We only need to consider the horizontal discretisation of the variables u, v and ζ, which are the eastward current, the northward current and the surface elevation, respectively. There are in fact three grids (that is, ways of discretising these variables in the horizontal) still in use, and these are shown diagrammatically in Figure 3.5; (a) is the Arakawa A grid, (b) is the Arakawa B grid, and (c) is the Arakawa C grid.

In the A grid, all variables are evaluated at the same location. At first sight this may seem logical, but bearing in mind that u and v are related to gradients of ζ, this turns out to be not very convenient. The B and C grids were developed so that points where the elevation was evaluated were always *between* points where the current was evaluated. This was first done by Arakawa, a meteorological modeller, in the mid-1960s. In the B grid, shown in Figure 3.5(b), both u and v are evaluated at the same point and the velocity points are situated at the point that is equidistant from the four nearest elevation points. In the C grid, this is not the case. Instead, the u points lie east and west of ζ points, and the v points lie north and south of ζ points. This is shown in Figure 3.5(c), and it is this Arakawa C grid that is most popular today with ocean and continental shelf modellers. However, the B grid does have the advantage of allowing a semi-implicit representation of the Coriolis terms and is also very useful for certain coarse grid schemes. Other grids are of course possible; in particular, it is more accurate to involve more values than just those on the neighbouring grid points. In practice it is found that much the same increase in accuracy can be achieved merely by decreasing the step size. This is also much more convenient than complicating the difference equations by the inclusion of many more terms.

In the vertical, a common device is to use σ-coordinates. These replace the standard z coordinate by

$$\sigma = \frac{z - \zeta}{h + \zeta},$$

so that the sea surface $z = \zeta$ is at $\sigma = 0$ and the sea bed $z = -h$ is at $\sigma = -1$. This way the domain of the problem (in the vertical) is flat at the top and bottom. The down side of using the σ-coordinate system is that derivatives of z have to be transformed into derivatives in terms of σ, which are more complicated. The sea bed and surface boundary conditions are, however, greatly simplified.

Chapter Four

Boundary conditions and validation

4.1 Introduction

We have seen from the last chapter that the motion of a sea or ocean is determined by physical laws. The laws of motion attributed to Sir Isaac Newton have for 300 years quantified the motion of physical objects. The sea is a fluid which is also governed by these laws. Together with the equation that makes sure mass is conserved, Newton's second law, translated to be valid for a fluid, is solved numerically by methods outlined in Chapter 3. The chemistry and biology of the sea is also determined by laws, however these laws lack the universal status enjoyed by the laws of motion. However, the methods of Chapter 3 can only be employed if it is possible to write down equations to discretise. At the edges of domains of interest, this is not possible without considering physically what is happening there. Different equations are valid depending on the character of the particular boundary. We shall see in this chapter that, far from being peripheral to modelling, these conditions that are literally on the edge or boundary are in fact central to the models and play a large part in determining the behaviour of the sea.

So, although the layman may believe that what happens on the boundary may not appear to be important, it actually drives the motion. It is all the more important, therefore, to understand precisely what is happening at the edges of regions. First, we need to distinguish between various types of boundary. Perhaps the most obvious boundary is the sea surface. It is also, unfortunately, one of the most complex. Another obvious boundary is the coast. The coast is a solid boundary, and the treatment of solid boundaries is certainly less controversial. Another solid boundary is the sea bed; but we need to be careful to know what we mean by solid. Sea bed boundary conditions have occupied modellers for a very long time, and some quite sophisticated models are now in common use. The reason for this close attention is not hard to fathom; it lies in the role the sea bed plays in providing a sink for momentum. There is far more contact between the sea and the bed than the sea and its coastline, and a great deal of effort has been devoted to engineering problems associated with the sea bed such as dredging, erosion and scour. Conditions imposed at the sea surface, the coast and the bed are called closed because there is a definite physical edge that dictates to some extent the type of equation valid at the boundary (relating the surface current to the wind perhaps, or imposing no flow through a coast). There are also open boundary conditions. These are not actual, physical boundary conditions, but arise because domains that are not closed lakes or the entire globe have to have edges that cross the open sea. Open boundary conditions are necessary to apply to the open edges of these models.

Another very different boundary condition arises from having time as a variable. All models have to start, and the state of the variables at time zero, the initial condition of the models is of course a boundary condition. In complex models that have significant non-linear terms, the start conditions are not important because the sea soon 'forgets' how it started to move and becomes, so to speak, wrapped up in its own dynamics. Systems or models with short memories are the type that lead to chaotic behaviour which is deterministic (that is, it lacks a random or stochastic element), but exhibits behaviour on many very different length scales. This is a popular topic nowadays, but the modelling of strongly non-linear systems is beyond the scope of this modest text. Having said this, in models that are not strongly non-linear, and that includes most of them, the start condition is very important. As examples, one can think of initial conditions that drive storm surge forecasts, diffusion models (see Chapter 7), and indeed weather forecasting itself, which would simply be a non-starter without accurate initial conditions.

Models begin by being simple and idealised so that the essential dynamics is present in glorious isolation, uncluttered by awkward boundaries or complex but unimportant effects. The rectangular ocean models of the next chapter provide particularly vivid examples of this. In very idealised models, the sides and bottom of the box-shaped ocean are flat, as is the surface. A rigid lid approximation is imposed whereby the sea surface is assumed solid. The open boundaries are simplified to be lines of flow (streamlines), and time does not feature. In spite of these idealisations, important features pertinent to the understanding of the general circulation of the ocean can still be predicted. Moreover, because of the elementary nature of the model, the user can see precisely what causes a phenomenon such as the western intensification of ocean currents.

As models are made more realistic, boundary conditions need to be treated with greater care. In particular, the different treatments of horizontal and vertical boundaries which stems from their differing scales are very important to maintain in ocean models. In models of less horizontal extent, such as an estuarial model, then the distinction between horizontal and vertical boundaries becomes less important. In an estuary, there are in fact only 'water–air' and 'water–solid' boundaries. Another recent development in modelling which fits naturally into a discussion of boundary conditions is data assimilation. Data assimilation is the incorporation of data into a model *as it runs*. It finds a natural home in limited area ocean models where a neighbouring model, or perhaps a larger one, can input information into the given model as it is running by having both models running in parallel. A second possibility is to input observational data into a model as it is running. For example, if a limited area ocean model is running, then as it evolves it could be possible for an eddy to migrate into the model even when the nature of the model makes the formation of such an eddy impossible. This would be done by using as boundary conditions the velocity and elevation appropriate to an eddy along one of the model's edges.

Another obvious use of data assimilation is in storm surge modelling (see Chapter 6). In storm surge models, it is wise to update the weather input as the model is running in order for the enhanced elevation to be predicted with the greatest possible accuracy. The increasing availability of satellite data makes the improvement of model prediction by data assimilation a real possibility in areas where it has not been in use so far. However, for this text, data assimilation takes a relatively minor role as its main use is in obtaining answers that fit observations, and not answers that improve the understanding of the fundamental processes. Let us now return to the main subject of this chapter and examine the different types of boundary condition more closely.

4.2 The sea surface boundary

The surface of the sea can be a flat calm, it can be mountainous, or any condition between the two. In ocean models it is commonly taken as flat, because over reasonably short periods of time, say a few minutes, the up and down movement averages to zero. Since ocean models are concerned primarily with bulk movements through ocean currents, time steps longer than this are used which implicitly implies averaging on the right time scale to eliminate the vertical movement of the sea surface. There is, of course, an entirely different vertical displacement of sea surface due to astronomical forces called the tides. These need special attention and form the subject of Section 6.2.

The surface of the ocean can thus be assumed flat (apart from tides). It can still move, however; it moves horizontally like an airport travellator. To force the ocean surface to act as if a solid barrier were against it is nowadays unacceptable. This was called the rigid lid approximation and flourished briefly in the 1950s and 1960s. It is much more usual to allow the wind to act on the sea and to move its surface tangentially through frictional forces. The momentum thus introduced into the surface is then transferred to the ocean underneath by the turbulence so created. A way to achieve this is to relate the velocity at the sea surface to the sea surface stress via a law which might be a stress rate of strain relationship (Newtonian eddy viscosity), or to use a more sophisticated law. These more complex laws are dealt with in the next section, and overall models that deal with coastal sea problems that require sea surface boundary conditions form part of Chapter 6.

It is also possible to allow for other boundary influences through the sea surface. If the temperature variation of the ocean is being modelled, the surface can be allowed to heat or cool due to outside influences (night and day, or the different seasons) and this can be incorporated by imposing temperature or temperature gradient conditions at the sea surface itself. A model would incorporate this via a source (or sink) term in the equation governing the diffusion of temperature downward through the water column. Other sea surface effects such as rainfall and evaporation have not been the concern of the modeller and will not be considered here, however they can be incorporated in principle using sea surface source or sink terms. In the newest ocean circulation models, the potential energy that arises from the rise and fall of the sea surface has an input into the salinity distribution. This is interesting in that models of the ocean circulation in the 1930s were entirely driven by the effects of precipitation, evaporation and freshwater inflow but these models were abandoned when the later wind-driven models seemed to be so much better at predicting the observed circulation. These latest models show us that we are wrong to reject completely such models. Instead, we must incorporate these effects alongside the more dominant wind driving in a more complete description of the surface (and coastal) boundary conditions.

4.3 The sea bed boundary

The most obvious characteristic of the sea bed is that it is a solid barrier and that water must not be allowed to pass through it. This might be obvious, but the smaller the scale of the model, the trickier such a criterion is to apply. A sea bed may first of all be steep, so that care needs to be taken that the boundary condition mentioned above involves the flow perpendicular to the sea bed. If the bed is flat, it may be sandy or muddy, in which case can it be assumed solid? Also, what about the cohesive properties (stickiness)? In a viscous fluid, all components of the velocity including those parallel to the sea bed, must be zero at the bed itself. Although the ocean is not a viscous fluid in the accepted sense, its equations are similar enough for such a boundary condition to hold. Finally, if very detailed modelling is to be considered, then some techniques from mechanical and aeronautical engineering modelling can be used. In particular, the defining of a roughness length to represent the character of the bed (sand, gravel or rocks), a laminar sub-layer where the flow regime is viscous but not turbulent, then a transition to turbulence when one is clear of the bed.

A quite recent successful modelling approach to the sea bed boundary has been to use so-called k–ε turbulence closure schemes. Although the models that utilise these schemes apply to the entire sea, it is the fact that this particularly accurate model is required to simulate turbulence in areas that are dominated by frictional effects which causes us to go into more detail here rather than in Chapter 6, which might be thought the more natural home for them. The k–ε models themselves were first developed by modellers of aeronautical systems where the ability to predict the detailed flow near critical parts of aeroplane wings was an essential requirement. In Chapter 2, we encountered the concept of eddy viscosity as a quantity that represented how stress is related to the rate of strain (or shear) of a turbulent flow. It is analogous to kinematic viscosity in laboratory viscous laminar flow, the principal differences being that turbulent viscosities are much larger than their laminar counterparts because turbulence is far more efficient at transferring momentum from one streamline to a parallel streamline, and also a turbulent eddy viscosity is not a fixed property of a fluid. (In laminar flow, the viscosity is as fixed a property of the fluid as is, for example density. In turbulent flow, eddy viscosity can vary from place to place and can change with time.) Eddy viscosities cannot, unfortunately, account for the observed behaviour of a flow adjacent to the sea bed. Instead, modellers of recent times have used the above-mentioned, more sophisticated k–ε turbulence closure models. What follows is a brief description of this more complex model of turbulence. By the very nature of this model, it is necessary to delve a little into mathematics. Those readers who are frightened by this can skip to the next section, or get what you can from the words between the symbols!

The rationale behind most turbulence closure schemes is to start with the well known assumption of eddy viscosity, but then to introduce other variables that can be related to length and velocity scales which represent scales of turbulence. To be specific, if K_q represents the eddy viscosity (or eddy diffusivity), then

$$K_q = lqS_a,$$

where l and q are appropriate length and velocity scales, respectively, and S_a is a factor that depends on the stability of the flow. This stability will, in turn, involve the density change with the vertical as well as the vertical shear of the flow, $|\partial u/\partial z|$. The velocity scale q and length scale l will themselves obey equations. For example, $\frac{1}{2}q^2$ represents the energy associated with the turbulence. In simpler models, S_a for example is simply a number and the complex structure of the turbulence model is carried in the two scales through the equations obeyed by them. It is quite usual these days for ostensibly simple two-dimensional models to have quite complicated turbulence closure schemes attached to them which can mimic successfully the momentum transfer just above the sea bed. This kind of model is often called '$2\frac{1}{2}$-dimensional', and has the additional merit of being much cheaper to run (and in some cases more reliable) than more complex fully three-dimensional models. One interesting feature of virtually all sea bed models that purport to be detailed models of dynamics is the presence of a layer adjacent to the bed where the velocity profile is logarithmic. That is, the speed U is related to the distance from the sea bed z by an expression such as

$$U = \left(\frac{\tau}{\kappa u_0}\right) \ln\left(\frac{z}{z_0}\right),$$

where τ is the sea bed stress, κ is a constant attributed to von Karman and usually given the value 0.4, and u_0 is a constant representative of the speed of flow just above the roughness elements. These roughness elements are the sand, rock, etc., that lie on the sea bed and interrupt the flow at the bed itself. The constant z_0 is a length that represents the average magnitude of the height of these roughness elements above the sea bed. At a height of z_0, the flow is no longer interrupted by sea bed debris. These ideas are displayed in Figure 4.1.

In addition to the correct modelling of the physics at the sea bed, at the bed itself, some approximations are usually required because numerical methods which incorporate grid boxes (see Chapter 3) are being employed. This means that one point

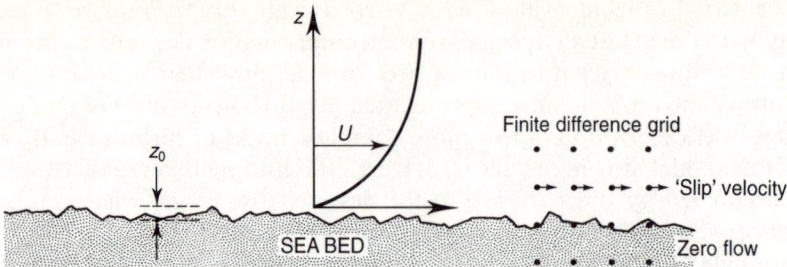

Figure 4.1 The sea bed, showing roughness length (z_0), finite difference grid and slip velocity.

will be above the sea bed, but the adjacent point must be in the bed itself. For this reason a *slip velocity*, whereby the velocity at the sea bed is not zero but simply some value, is a common proposal. The value chosen is consistent with an appropriate quadratic or linear friction law that relates frictional stress directly to velocity at the bed. The usual controversy over whether to use such a law or whether to use a no-slip condition is sterile because of the numerical approximations (see the notional grid displayed in Figure 4.1). In fact, the numerical application of a no-slip law at the sea bed itself is precisely equivalent to the imposition of a linear friction law a little above the bed at the nearest grid point. Having said all this, however, sophisticated turbulence closure schemes are now central to modelling the dissipation of momentum at the sea bed correctly, and numerical approximations of the sea bed boundary conditions merely alter the actual points at which these conditions are applied, not their application.

4.4 Coastal boundary conditions

In the ocean, the coast and the edge of the continental shelf can be taken as being virtually synonymous. This is because of the magnitude of the horizontal length scales. If a model is to cover the entire horizontal extent of an ocean, perhaps 10 000 km, with 100 points, then each point has to be 100 km apart. This then means that a typical continental shelf is entirely lost between the coast and the first grid point. The coast itself, under these circumstances, can be assumed to be a shear vertical cliff. Exactly similar representations of the slip or no-slip boundary conditions at the sea bed are possible at the coast too when it is a vertical cliff. However, there is a subtlety that arises because the Coriolis parameter varies with latitude which makes a no-slip condition dynamically distinct from a formulation which includes a slip velocity. There is no need, of course, for the complicated turbulence closure schemes in these large-scale models. We take a closer look at coastal boundaries in the next chapter since this is the natural place to consider the change of Coriolis parameter with latitude.

For smaller-scale models, the continental shelf must be taken into account. Normally models do not straddle the continental slope; they either lie entirely on it or it provides the location for boundary conditions of the deep sea models. Models of the continental shelf itself have open boundary conditions which need careful handling and deserve a section of their own.

At smaller scales still, say coastal inlets or estuaries, the coasts are treated in various ways. Often they are steep sided and a vertical face is allowable; on other occasions there may be mud flats or sandy beaches of minimal slope which demand that the depth of the water itself is zero. There is a potentially serious problem if the sea dries inside a model region, because the depth occurs as a parameter in the denominator of various terms of the governing equations. One approach is to allow the sea to dry in cells in such a way as to compute the velocity as zero before (potentially) dividing by the zero depth. Research is still active, however, on the effects this has on the velocities elsewhere in the model domain. When there are drying regions present in a model, the horizontal domain of the problem changes with time (perhaps due to tides). Such a formulation is difficult to solve since the shape of the coast becomes an unknown of the problem. Dealing with moving coastlines can be technically difficult, but as a concept it, as it were, holds water. In Chapter 6, more is said about models of continental shelf seas and what boundary conditions are appropriate.

4.5 Open boundary conditions

The boundary conditions considered so far can be considered *natural* in that they arise from a sea being surrounded by solid or air. It is the physics of the interface that by and large dictates the form that the boundary conditions take. Open boundary conditions occur simply because a model must end where there is no coast. Typical open boundaries are the edge of the continental shelf; Figure 4.2 shows a typical domain off northwest Europe.

In this example, there is an extensive open boundary which more or less marks the European continental shelf. Estuarial models and models of coastal inlets will have open boundaries with the sea. Even deep ocean models which span the Atlantic or Pacific Ocean from west to east often have equatorial and northern open boundaries. In this latter case, the open boundary is dealt with similarly to a solid boundary in that it is assumed to be a streamline. (No flow passes through a streamline, although in fact all that is required is that there be no *net* flow; there can be some transfer of fluid, but it has to be the same in each direction.) For smaller models, however, the open boundary is treated differently. The essential feature a modeller wishes to retain is a lack of sensitivity as to where precisely the boundary is actually placed. If this is the case, then some confidence can be placed on the solution. Open boundaries are often subject to 'radiation conditions'. This is a way of allowing the open boundary to 'let out' flow, etc., from the domain without reflecting any energy or momentum back into the domain. Numerically, this is hard to achieve exactly, but the further a boundary is away from the region of interest, the better (more

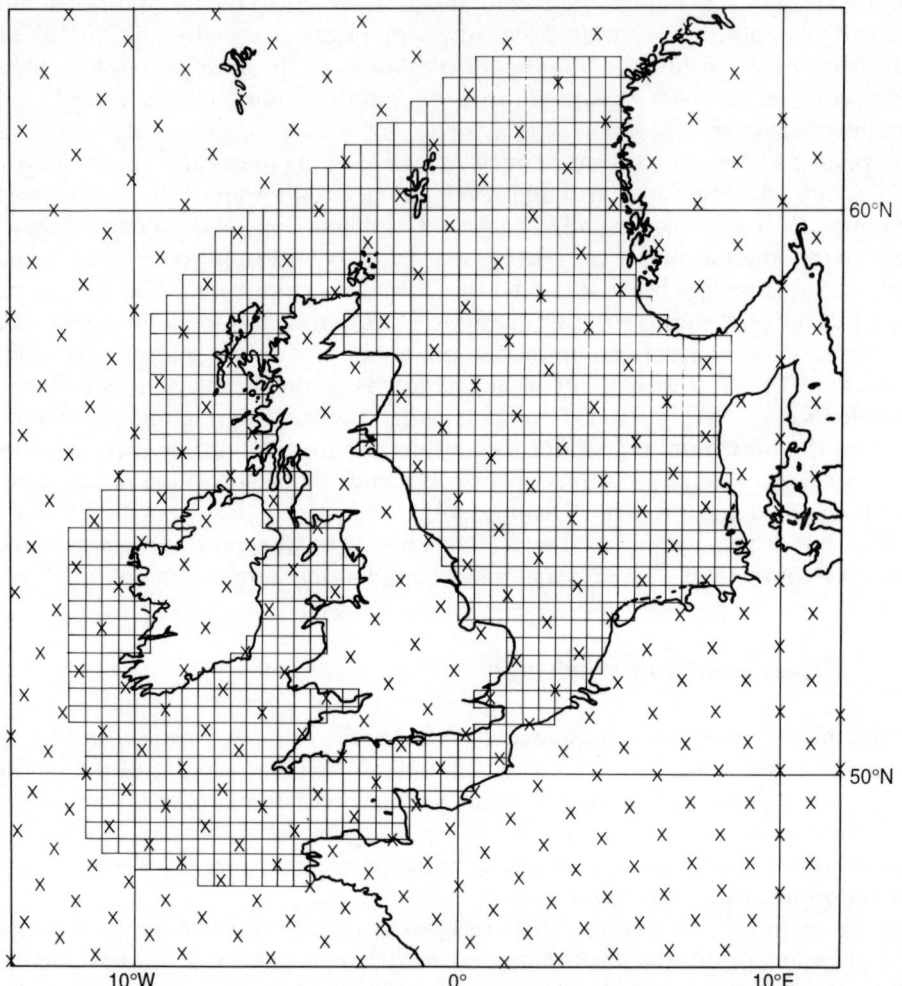

Figure 4.2 The continental shelf sea model (CSM) used as the basis for the IOS surge forecasts, with grid points (×) of the Meteorological Office 10-level weather prediction model which supplies the required forecast winds and atmospheric pressures. From Flather (1979). Reproduced with permission.

accurate) the results. Another way is to 'nest' models such that a fine grid model is embedded inside a coarser grid model. Much care needs to be taken with this, but it is a preferred solution to simply extending the grid until it is 'far enough away'. For no matter how far away an open boundary, its effect can still travel through the entire domain well before the currents have settled down. In fact the effects due to the boundary have to attenuate, and this implies the presence of some kind of friction which may not be consistent with the physics of the sea.

4.6   Model validation

The validation of a model is the comparison of model output with what can be termed current knowledge. This knowledge usually comes from observation (see Chapter 1). It might seem as if nothing can be easier than to compare the output of a model with observations which usually occur in the form of data. However, this is not the case except perhaps for tides where there are very long and very accurate records (in most instances). If a model is run, and values of the variables are obtained, one is faced with the question of how accurate these results are. To take a specific example, a domain may be overlaid with a two-dimensional grid (Figure 4.3 is typical) and values of surface elevation, eastward and northward velocity obtained as model output at all grid points. How good are the results? Perhaps an observational programme over the same area or a convenient cruise has occurred and instruments have produced some data. First, the data will not have been taken at precisely the same locations as the model grid points; second there will in all likelihood not be enough data; and third not all of the data will have been collected at the same time. All of these factors make comparison between data and model output difficult. In order for some values to be compared directly, interpolation has to take place. Interpolation is a numerical technique that enables values of a variable at some intermediate location to be computed from those that surround it. It is a generalisation of curve fitting and there is now quite an extensive library of software that can help with this process (terms associated with interpolation include splines and least squares, both of which are useful, and Lagrange interpolation, which should be avoided). This then deals with one problem, but if there are not enough data, or they were recorded at the wrong time, there is not much we can do. Satellite obtained data promise to be very useful here and could provide a leap in the sophistication of model validation. Suppose that we have adequate data, what do we do about comparing these data with the output from the model? The easiest thing to do is to 'eyeball' both and assess the significance of any differences. This is still often the only method of validation used in marine science. While not excusing this very unscientific methodology, it is perhaps understandable that oceanographers and ecosystem modellers should be unwilling to expend large amounts of time and energy on sophisticated statistical techniques when much of the data are so roughly hewn from the sea. Nevertheless, there are techniques that can be of some use. If a direct comparison between model output and observations is possible, then one can analyse the differences between them statistically. For example, using the statistics of sampling, it is possible to tell whether differences are significant by using the t-distribution and to place confidence limits on the significance of these differences. If, further, it is possible to postulate that, as a *null hypothesis*, there should be agreement between observations and model output, then one can define this measure of agreement via the χ^2 test. This is a statistic that is defined by the expression

$$\chi^2 = \sum \frac{(\text{Observed} - \text{Expected})^2}{\text{Expected}}.$$

The 'observed' values are the model results, and the 'expected' values are the corresponding observations. The Σ sign denotes summation over all data points. Of course, one could swap the role of model results and observations; this would in theory 'test' the observations against the model, assuming the model to be correct. If both are reliable, then there is no problem. If neither is reliable, then we get the standard arguments between those who measure and do fieldwork, and those who model and do calculations. Once the χ^2 statistic has been calculated, it is a simple procedure to look at a table (see for example, Murdoch and Barnes (1974) *Statistical Tables*) and ascertain whether or not departures are significant. Of course, it is tempting to regard this as a definitive argument for or against a particular model. In reality, standard statistical tables have built into them certain assumptions involving the normal distribution that particular observations or model output may disobey. As a general rule, it is most unwise to use statistical methods that are more complicated than is warranted by the veracity of the data! If in doubt, contact a professional statistician. A necessarily very brief introduction to some statistics is given in Chapter 9.

4.7 Examples of representing boundaries

The final boundary condition to consider is the start conditions. There is very little to say here, except that it is good practice to get as accurate data as possible to start a model. This is particularly true for storm surge models and ocean models that are not strongly non-linear. However, there is a class of models (tidal models spring to mind) where the object of the exercise is to run the model in a predictive state to be as accurate as possible and *independent* of start conditions as soon after the start of the model as possible. For this kind of model, it is common to start the model by assuming a flat stationary sea and to allow the open boundary to input motion in order to *spin-up* the sea itself.

In the remainder of this section, we revisit the two examples of the last chapter, look at the boundary conditions, and determine the role they play in the subsequent solution of the models.

4.7.1 Finite difference schemes

The model examined in Chapter 3 was a semi-implicit model. However, this look at the edges is virtually independent of this kind of detail. Let us look at how each kind of boundary condition is dealt with in models that make extensive use of finite difference schemes. At a solid boundary, whether it be the sea bed or a coast, the problem is essentially that a grid point that is in the sea and at which an equation is valid needs to be discretised in terms of differences, but these differences need to be calculated using values that are in dry land because that is where the adjacent point is. The point in question then has to be rejected insofar as being able to write down discretised versions of the governing equations is concerned. Instead, the

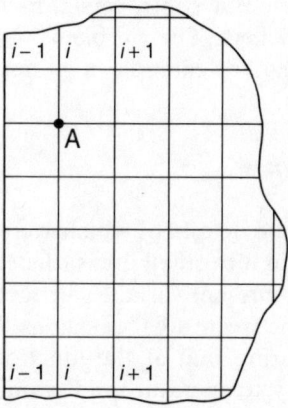

Figure 4.3 The finite difference grid near a boundary.

boundary conditions need to be examined, and to be discretised appropriately. As an example, suppose that there is a straight solid boundary as shown in Figure 4.3. At the point marked A, which is in the interior of the domain and therefore in the sea where various equations are indeed valid, the boundary condition that there is no flow through the boundary is discretised in the following way:

$$\frac{\partial U}{\partial x} = \frac{U_{i+1} - U_{i-1}}{2\Delta x} = 0.$$

This replaces the full discretisation of the equations. Also, because U_{i-1} is due to be evaluated on dry land, it is taken as zero. Zero flow through the boundary therefore automatically leads to a zero value for U_{i+1} as well. This is obviously rather unrealistic, and various ways round this have been proposed, such as to stagger the grid such that velocity points are never actually on coasts, and to make a fictitious extension of the grid one point into the coast so that the domain is not artificially shrunk. The Arakawa C grid, outlined in Section 3.4.3, is one modern finite difference grid that facilitates this.

At the sea bed, a similar problem can occur. The specification of the velocity just above the sea bed rather than actually on it is tantamount to specifying a slip velocity at the bed itself, simply because the resolution of the finite difference scheme does not allow them to be distinguished. The way out of this is to use something other than layers in the vertical, or at least to use a continuous representation of velocity close to the bed itself. If finite differences are used, then the sea bed boundary condition typically becomes the specification of stress at some point above the bed. Practically, however, this is not a problem since our observational colleagues have to measure the stress at some point above the sea bed and not actually on it. The standard height above the bed is 100 cm. It is then this measurement that is used as the boundary condition, and this is of course not controversial as it marries what is available to what is required by the modellers. The problem is that neither precisely matches reality, so that validation using, for example, a χ^2 test would not be valid.

4.7.2   Finite element schemes

In finite element schemes, some details of which were given in the last chapter, the domain of a problem is divided into small areas (the elements), usually triangles, and simple functions are used to represent variables in each of the small areas. There are, of course, internal borders between adjacent areas, but a good representation of reality is maintained by ensuring that at the junctions between the areas physical quantities such as fluid velocity, temperature and salinity are free of unphysical sudden jumps. At coasts, however, things are different. The principle behind the finite element method is that, if the entire domain is considered – that is, the flow, temperature, etc., are each examined over the whole sea – then conditions can be imposed that correspond to basic physical laws such as the conservation of momentum and mass. This is done by forming a sum over all the elements and then applying each physical law. Boundary conditions such as no flow through coasts or a specified tidal amplitude have no place, directly, in such a scheme. In fact, what happens is this. The equations that arise from the imposition of these laws over all the elements lead to enough equations to be able to determine every simple function in each small area. In fact, it turns out that there are *too many* equations because the numbering of the locations (nodes) where the variables need to be determined does not distinguish between which is an internal node and which is actually on the boundary. This problem is overcome by writing down *all* equations (even those that are false due to the nodes being on the edge of the domain), then deleting the false equations and replacing them by the appropriate boundary conditions written in terms of the functions valid in these border elements.

Technically, the finite element representation of (say) tides is found by inverting a large matrix corresponding to solving a large number of simultaneous equations. Typically, there are 1000 equations in 1000 unknowns. Most of these equations contain only four or so unknowns, corresponding physically to any given unknown being dependent only on those variables in its immediate vicinity. The matrix is therefore largely full of zeros with non-zero entries clustered around the main diagonal. Mathematicians call these banded matrices, and there are special methods for dealing with them efficiently. The introduction of boundary conditions may lead to isolated off-diagonal entries which are potentially a nuisance, but not fatally so since judicious row operations usually restore the banded nature of the large matrix. These days there are inbuilt routines that generate finite element meshes and have inside them ways of incorporating different types of boundary conditions. Those interested in the technical detail should consult a specialist text such as Mitchell and Wait (1985).

For the oceanographic modeller, the most convenient aspect is that the boundary conditions are embedded in the equations themselves and are not a separate feature, as is the case with finite differences. Of course there is a temptation to think of finite elements as in some way 'more natural' because of this, but this is a dangerous illusion, since both are as good or as bad as each other!

Validation studies have taken place for finite element models. For the study of Westerink *et al.* (1994) outlined in the last chapter, the output of the model was compared directly with well known measured tidal constituents. In this model, a proportional standard deviation was defined as follows:

$$E = \left(\frac{\sum_{l=1}^{L} (\zeta^c(x_l, y_l) - \zeta^m(x_l, y_l))^2}{\sum_{l=1}^{L} (\zeta^m(x_l, y_l))^2} \right)^{1/2},$$

where the symbols have the following meanings:

L = total number of elevation stations within a given region;

(x_l, y_l) = the location of the elevation station;

$\zeta^c(x_l, y_l)$ = the computed elevation amplitude at a given station;

$\zeta^m(x_l, y_l)$ = the measured elevation amplitude at a given station.

In the model, a number of tidal constituents were simulated. In particular, the average error as measured by E was between 18.2% and 45.3%, where eight tidal constituents were considered and the entire domain was covered. Nothing statistical was attempted, but the regions of greatest error coincided with the amphidromic points (where the tide vanishes). Also, those stations with poor convergence properties (i.e. the model elevation only slowly converging to an answer) gave the poorest comparison with measurement.

For wind-driven flow a statistical comparison between model output and observational data is hardly ever done since observations usually drive the model, rendering them dependent and rendering statistical approaches invalid. The future availability of satellite data could herald the onset of proper validation studies. For the model of Westerink *et al.*, not only were validation tests in the form outlined above carried out, but sensitivity tests were also done. This took the form of splitting elements and examining the subsequent changes in both the amplitude and phase of the tide.

Chapter Five

Large-scale ocean dynamics

5.1 Introduction

In this chapter we shall be looking at the dynamics of the ocean on an ocean-wide scale. This is historically the first part of the ocean to be modelled, dating back to the 1930s, when the first tentative modelling steps were taken, to the late 1940s and early 1950s and the more assured work of H.U. Sverdrup, H. Stommel and W.H. Munk. These pioneering modellers were the first to recognise that the overall shape of the steady currents in an ocean such as the Atlantic were due to simple balances. The bulk of the ocean reflected a geostrophic balance between Coriolis acceleration and pressure gradient forces in a parallel fashion to the balance in the atmosphere. Superimposed on this are flows induced by the variation of Coriolis acceleration with latitude. At the equator where a tangent plane to the Earth is parallel to the Earth's axis, there is no component of the Earth's rotation in the direction perpendicular to the Earth's surface. On the other hand, at the pole, a tangent plane (*the* tangent plane) to the Earth has the axis of rotation passing perpendicularly through it. Elsewhere on the Earth, a tangent plane has differing amounts of rotation. We talk of tangent planes because we are discussing rotation; we really are only concerned about points in the ocean which correspond to particular latitudes and longitudes, but points have zero horizontal dimensions and so are not convenient indicators of the Earth's rotation. Latitude is of prime concern, since once this is fixed one is on a circle on which all points (or tangent planes) experience the same rotation. As latitude changes, so the amount of rotation experienced by the tangent plane changes. It is this change in rotation (strictly, change in the vertical component of Coriolis acceleration) that gives rise to the phenomena of western boundary currents – the Gulf Stream, the Kuroshio, the Agulhas current, the Brazil current, etc. Simple models required idealised geometry, especially idealised straight coastlines and flat ocean bottoms. Once these were discarded in the 1950s, numerical methods were employed which heightened the realism of the models' outputs. However, numerical techniques were employed at this stage merely as means of evaluating formulae. In the later decades of this century, more sophisticated numerical models have been employed which have enabled the modeller to portray realistic internal dynamics. In particular, modellers of the ocean circulation are now turning their attention to important but slightly smaller-scale phenomena such as the influence of local topography on ocean currents and the seasonal changes of particular currents such as the Agulhas' response to transitional eddies. Only now has this kind of detailed modelling become possible. Previously, only observations and some descriptive physics could help us understand such local events. The verification and validation of these more complex models have been greatly aided by the large international observational exercises. All this ensures that today's models of how large-scale processes feed their dynamics into smaller-scale features are among the best models in all oceanography.

5.2 A perspective on global ocean modelling

In this section, we focus on the historical development of ocean circulation models. Not only are these interesting in themselves, but they represent a splendid example of the art of modelling in that a very complicated display of physics (the movement of the ocean) is reduced to some simple balances that nevertheless explain salient oceanic features. There are several key papers that chart the progress of our knowledge of how the ocean moves. The famous paper of Ekman (1905) which first gave insight into how the wind drives the upper layers of the ocean, is perhaps an obvious starting point. Before this, most models that had application to the sea were primarily mathematical; for example, the Victorian papers of Hough on spherical harmonics, and before that papers by Laplace dating back to 1776 on global tides. However, let us concentrate on twentieth-century developments.

Between 1905 and the Second World War, no significant advances in ocean modelling were made. That is not to say that there were no papers of relevance, for in 1933 Goldsborough sought to explain the movement of the ocean by examining the evaporation and precipitation on a global scale with some success, and in 1938 Carl Gustav Rossby published the paper that gave the upper atmospheric waves their name. These waves are also found in the sea, but their existence in the oceans was not quantified until the 1950s and 1960s.

The major modelling advance was made in the late 1940s in quick succession by Sverdrup, Stommel and Munk. Figure 5.1 is reproduced from Stommel's 1948 paper. It highlights the importance of the variation of the Coriolis acceleration (strictly, the vertical component of Coriolis acceleration) with the latitude. Without this effect, the circulation due to the wind forcing shown, which incidentally is supposed to represent averaged trade winds and westerlies, is entirely symmetric – a symmetric gyre, to use the correct phrase. Once the model includes latitudinal variation of the Coriolis acceleration, an intense western jet results from precisely the same wind and boundary conditions. Referring back to Chapter 1, the modelling process, and in particular the flowchart in Figure 1.1, it can be seen that the first model might be one that assumes that the Coriolis parameter is a constant. The equations are solved and the results are compared with observations. Once this is done the inadequacies are obvious for all to see, for there is a symmetric gyre with no sign of western intensification. The

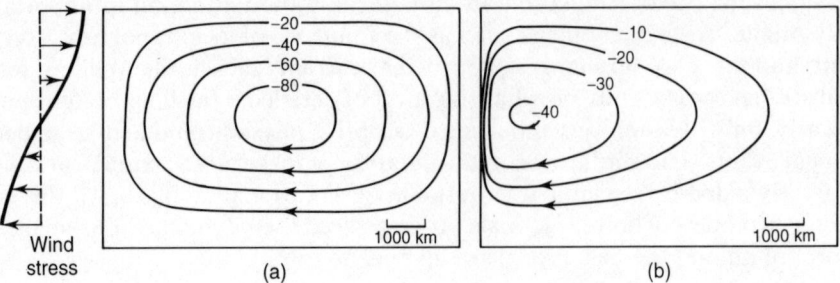

Figure 5.1 Flow patterns (streamlines) for simplified wind-driven circulation with: (a) Coriolis force zero or constant; (b) Coriolis force increasing linearly with latitude. From Stommel (1948). Reprinted with permission.

flowchart tells us to return to the assumptions box and revise them. Stommel's breakthrough was to include the effects of the changes in the Coriolis parameter with latitude. This then becomes the second model, and the subsequent solution of this new set of equations leads to a much more realistic solution that does include western intensification. This second model is thus acceptable. We are thus certain of the effect that causes the western intensification of ocean currents, so we can explain the existence of the Gulf Stream in the North Atlantic, the Kuroshio (Kiwo-Siwo) in the North Pacific, the Brazil current in the South Atlantic and other western boundary currents. Once Stommel had successfully isolated the variation of Coriolis acceleration with latitude as the sole cause of the western intensification of ocean currents, fluid dynamicists came up with vorticity conservation arguments which neatly explained all. Hindsight is a wonderful thing. Munk's contribution (1950) was to build on Stommel's work to include more realistic friction, more realistic wind forcing and (slightly) more realistic shapes for ocean basins.

It is not obvious why the variation of Coriolis parameter with latitude should play such an important role in the western intensification of ocean currents. Some insight can be gained without detailed knowledge of fluid mechanics, as follows. First, we need an axis which is *outside* the Earth (technically, we need an inertial frame of reference, but a single axis serves our purpose admirably). Without loss of generality we can assume that this axis is parallel to the north–south axis of the Earth since this does not move as the Earth rotates. Imagine a plane that is a tangent to the Earth at a particular point. The important feature of this point is its latitude; it does not matter about the longitude. A tangent plane will, naturally enough, rotate with the Earth. If the plane is at the North Pole, it rotates about the Earth's axis, making one complete revolution each day. On the other hand, a plane at the equator will rotate in a very different manner. In fact, if you imagine yourself standing on this plane facing north, as the Earth rotates you will continue to face north, although, crucially, this direction is the *fixed* direction of the axis placed outside the influence of the Earth and its rotation. This tangent plane therefore does not rotate as far as objects standing on it are concerned. At an intermediate latitude, the fact that you, standing on the plane, always face north as the Earth rotates means that you do have some rotation, but that that rotation is angled to the fixed axis. This angle is in fact the co-latitude, that is, 90° minus the latitude. The rotation we are attempting to focus on here is the vertical component of the Earth's rotation; this is the important quantity in ocean dynamics. This aspect of rotation which is zero at the equator and maximum at the North Pole can be taken as representative of a quantity called *planetary vorticity*. South of the equator, it reverses sign, but it remains true that the further from the equator one is, the greater the absolute value of the planetary vorticity. Now, vorticity is one of those quantities that likes to remain constant in a particular current. (For those familiar with mechanics, energy is similarly conserved under some circumstances, although there is a much clearer analogy between vorticity and the mechanics quantity angular momentum, which is less well known outside technical circles.) Thus as a current travels north from equatorial regions as, for example, does the Gulf Stream, it gains planetary vorticity from the increase in the vertical component of the Earth's rotation rate. It therefore has to lose an equivalent amount of vorticity from another source. What happens is that the internal fluid vorticity decreases until all available has been transferred to planetary vorticity (or until other circumstances change the dynamics). It is at this point that the Gulf Stream crosses the Atlantic. In reality, of course, there are further complications that render close scrutiny of this picture inappropriate for the detailed description of the behaviour of the Gulf Stream; for example, the role of friction (read turbulent eddy diffusion). If an intense current is adjacent to a coast, a *shear* is produced. If this current is flowing north, in the northern hemisphere on the western side of the ocean, then this shear is positive. A current as intense as the Gulf Stream produces a shear that is responsible for the bulk of the vorticity in this part of the northern Atlantic. It is the interchange of vorticity between planetary vorticity and internal fluid vorticity that explains the western intensification, although we have not done enough basic fluid mechanics to display this convincingly here. Instead, the reader is referred to graduate texts such as Pedlosky's *Geophysical Fluid Dynamics* (1987).

The next section will take us through more recent advances in large-scale ocean modelling, including other effects such as overturning through temperature and salinity differences. We shall also take a look at equatorial ocean models and the role played by large cooperative exercises, some of which are continuing and are seen as the way forward in advancing our knowledge on this scale of ocean dynamics.

5.3 Steady ocean circulation models

Let us start with a description of those features included by Stommel in his 1948 model of circulation. His ocean was square, it had straight north–south coastlines at both east and west edges, together with streamlines at the north and south boundaries. Friction was included as proportional to ocean current. The physical interpretation of this is that the ocean circulation is assumed to be two-dimensional or, equivalently, we concern ourselves only with a depth-averaged current. In this case, the friction is due to dissipation at the sea bed. With the wind inputting momentum via the sea surface, and friction causing momentum to be taken out (at the sea bed in reality, but because of the two-dimensional nature of the model this detail is lost), a balance is possible and a steady gyre is created. There are no non-linear advective terms, and the pressure gradient terms are unimportant. The balance is purely between Coriolis acceleration and friction, and in idealised geometry with a square ocean with constant depth. That a model as idealised as this gives great insight is a testament both to modelling and to the abilities of Henry Stommel. Many more details of his model(s) can be found in his fascinating book, *The Gulf Stream* (1965).

Munk's model, taken in its simplest form can also apply to the same square ocean with the same idealised wind forcing and boundaries. Crucially, however, Munk's model represented friction via the more usual terms associated with turbulent horizontal eddy viscosity. This implied that the structure of the current could cope with a no-slip condition at the north–south boundaries. Rather unexpectedly, this, and this alone, resulted in the model predicting not only the now familiar and expected western boundary current, but adjacent to this, and seaward of it, an opposing weaker countercurrent. The fact that such a countercurrent is observed is evidence for Munk's representation of horizontal momentum transfer and against the rather simple friction law of Stommel's model. Once more, the merit of a simple model is seen clearly. The beautiful uncluttered model of Munk and the equally beautiful uncluttered model of Stommel are easily compared, and it can be quickly deduced that a countercurrent is solely the result of the different representations of friction in the two models. Again this can be put into the framework of the modelling process flowchart of Figure 1.1, where this time the first model (Stommel's) is rejected on the grounds that it does not predict the observed countercurrents. The second model with a more complicated representation of the turbulent diffusion does predict these countercurrents and thus becomes acceptable.

All of these early models neglected any variation of current, temperature or salinity with depth. Later models have incorporated some depth-dependent characteristics and have consequently been more complex. This complexity has tended to obscure some of the underlying features, this is due principally to the unavoidable introduction of a multiplicity of length scales. These new length scales arise as a direct consequence of introducing depth-dependent quantities. Models that have variables that depend on the vertical coordinate are called *baroclinic*, and the depth-independent ones are called *barotropic*. These rather strange names actually mean something rather technical in terms of the internal thermodynamics of the ocean, but fortunately the interpretation that barotropic means no depth dependence, and baroclinic means possessing variables that depend on the vertical coordinate (that is, having vertical structure) is quite adequate for most purposes. It is recognised that a current such as the Gulf Stream is largely confined to the upper part of the ocean, that is, to that part of the ocean usually called the mixed layer. It is this layer that experiences the direct influence of the weather, in particular the wind that is to a large extent responsible for the existence of the Gulf Stream. The confinement of this current, and other similar currents, to the mixed layer should therefore come as little surprise. However, we

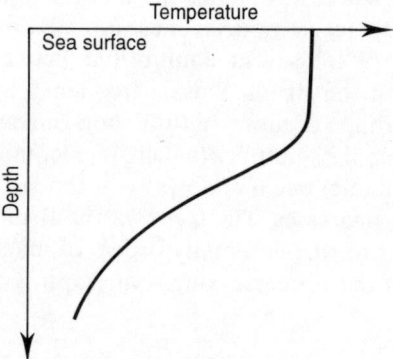

Figure 5.2 A typical temperature profile with depth.

need to ask what are the modelling implications of this. It is tempting to equate the mixed layer to the Ekman layer where the balance is between the Coriolis acceleration and the vertical transfer of momentum (the vertical eddy viscosity term in the simplest representations). This is sometimes true, but unfortunately not always. The ingredient that is ignored in Ekman dynamics is buoyancy which, on the kind of scales being discussed here, is almost always due to the change in temperature with depth (see Figure 5.2).

If there is a surface current shear of the type generated by Ekman balances in the surface layer of the ocean, then as this water tries to mix with that immediately beneath it, there is often a buoyancy force which tries to prevent any such entrainment. The dimensionless number that governs this process is the gradient Richardson number (see Section 2.4). Now, consideration of this kind of dynamics often means the incorporation of smaller scales into any hitherto depth-independent model, for models of entrainment have a respectable laboratory history and occur on scales where the modeller is untroubled by the Coriolis parameter. Obviously, here we are concerned with modelling oceanic structures that must be directly influenced by the Coriolis acceleration, and the appropriate modelling strategy is to include buoyancy forces in our model. This is done by writing down explicitly equations that govern the distribution of temperature in the ocean, and this in turn results in many more modelling choices as far as terms to compare for relative importance. An alternative strategy is to consider the ocean as consisting of two layers of slightly different density, one above the other. This gives rise, not to a Brunt–Väisälä frequency but to an analogous parameter $g(\rho_1 - \rho_2)/h_n\rho_1$. Here, h_n is a representative depth, commonly taken as $h_1 h_2/(h_1 + h_2)$, where h_1 and h_2 are the two layer depths, and ρ_1 and ρ_2 are the densities in the upper and lower layers, respectively. Most large-scale models that have followed either of these routes have used a non-dimensional set of equations which lead to non-dimensional parameters involving the nature of the stratification (that is, the variation of temperature with depth). The commonest parameters are the

Prandtl number, $\kappa_v/(fD^2)$, where κ_v denotes the thermal equivalent of eddy viscosity which is useful in large-scale steady models. These models are concerned with balances which are representative of long-term equilibrium between diffusive effects. The parameter N/f, the ratio of the Brunt–Väisälä frequency to the Coriolis parameter (itself a frequency), is perhaps a more natural non-dimensional representation of stratification. This dimensionless number (which, incidentally, does not seem to have any universally accepted name) occurs naturally in models that are concerned with evolutionary or wave-like processes. The fact that the Brunt–Väisälä frequency N is seldom constant and can run through many orders of magnitude begins to indicate some of the complexity of the processes one is attempting to model.

5.4 The Gulf Stream and other large-scale phenomena

The Gulf Stream is perhaps the best known of oceanic phenomena. The intense western boundary current which travels up the Florida coast, to leave the USA at Cape Hatteras and cross the Atlantic has a large part to play in the warmth of the winters in the UK, Norway and the rest of coastal Western Europe. During the 1950s and 1960s models were developed that began to explain some of its details. One question was why the Gulf Stream leaves the coast of the USA precisely where it does. (We now know how to explain this, at least qualitatively, in terms of the conservation of planetary vorticity. As the Gulf Stream travels north, it loses its internal fluid vorticity to planetary vorticity until a latitude is reached where there is none left to transfer. At this point it cannot travel further north, so it has to turn to cross the ocean.) Another question was what causes the variability of the Gulf Stream, both in terms of meanders as it crosses the Atlantic, and in terms of variations in intensity. Despite wide recognition of this variability, there is no question that the Gulf Stream, along with most other western boundary currents are permanent features of our oceans. They are always there, although of course they may vary in intensity and they may vary slightly in position. Since the main driving force is the wind, and the wind is extremely variable, one has to deduce that the Gulf Stream and indeed most other western boundary currents are caused not by local conditions but more by some kind of average climatic conditions. It is this fact that is primarily responsible for the success of the early models, for the driving force behind these models is the climatic average wind, not the fluctuating day-to-day wind. It turns out that this was yet another very good modelling assumption, at least if the attention was focused on the Gulf Stream which, if not precisely steady, certainly has a permanence which leads one correctly to a driving force that has a predominantly steady character. In more recent times, attention has focused on the variability and features associated with this variability, such as the wavelength of the meanders and the offshoots of the Stream itself which often form Gulf Stream rings. As far as detailed modelling of the Gulf Stream itself is concerned, the main recent advances have been in the representation of vertical mixing.

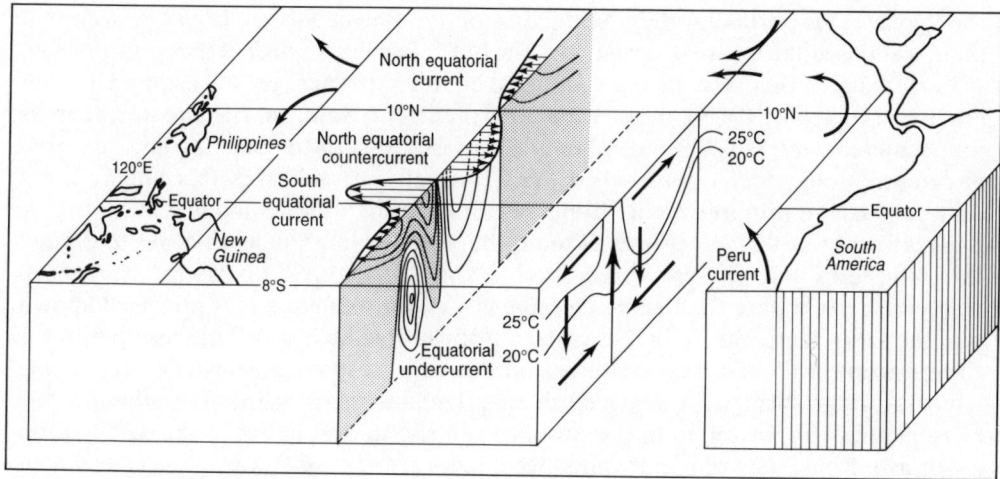

Figure 5.3 Schematic diagram of the horizontal and vertical circulation in the tropical Pacific Ocean. From Philander (1990). Reproduced with permission.

In contrast with the predominantly steady Gulf Stream, the Somali current is extremely variable and obviously directly responds to the onset of the monsoon. It is therefore principally *locally* driven. In spite of or perhaps because of this, the Somali current represents the largest flow of all western boundary currents; this must therefore be due to the fierce but predictable monsoon winds that whip up the ocean surface around the Indian subcontinent every summer.

At the equator there are currents that run east–west. The mechanism that causes these currents is principally the change in sign of the Coriolis parameter at the equator, although there are other effects such as the heat balance, and the physical structure of the oceans themselves (for example, the Pacific Ocean has a much larger fetch than the Atlantic Ocean in equatorial latitudes). Nor is it just a question of the currents being there to 'complete the gyres' by recirculating the eastern water back to the west to be picked up by the western boundary currents, for there are also countercurrents on either side of the principal equatorial currents. These are most well formed in the Pacific which has the longest fetch. Related to the dynamics of the equatorial current systems is the phenomenon of equatorial upwelling, whereby water is upwelled from below the thermocline to the surface. Coastal upwelling is perhaps more interesting, not least because of its vast economic importance to countries such as Peru and Nigeria. There is a close relationship between this coastal upwelling and the strength of the equatorial countercurrent, particularly in the southern Pacific where the upwelling at the Peruvian coast ceases from time to time (anything between three and eleven years), and is termed *El Niño* (see Section 5.7.2). In recent years a concentrated effort has been made to understanding the details of the equatorial circulation. Much of this effort has focused on the transient phenomena, the most obvious being *El Niño*. However, even the mean circulation (see Figure 5.3) is complex. This particular diagram focuses on the Pacific Ocean. It will be seen that there is at the equator a west to east flow, the bulk of which is north of the equator itself.

Further north there is a strong equatorial countercurrent (west to east) and, further north still (north of 10°N) another equatorial current. South of the equator, there is no countercurrent as such, but there is a strong sub-surface current (the equatorial undercurrent, in older books called the Cromwell current). It is this undercurrent that has such an important part to play in El Niño. In a modelling text such as this, it is pertinent to ask what processes are to be included in a model in order to predict this long-term but rather complicated current structure. No doubt, the mean temperature field over the Pacific Ocean has a major influence. It is now well known that the mean temperature in the western tropical Pacific is much higher than in the eastern tropical Pacific. Superimposed on this east–west asymmetry is a very strong temperature gradient with depth, with its attendant thermocline. This thermocline has a depth of about 150 m in the west but can rise to the surface in the east to form a surface front between erstwhile deep, cool water and warm surface water. Superimposed on this thermal forcing is the ever-present wind stress forcing. It is important to realise that the thermal asymmetry is there in the *mean* temperature field, it is not a merely transient feature. If one does focus on transient currents, then both the fluctuating wind stress field and the variations in temperature with time need to be taken into consideration for an accurate picture of the forcing. It is now realised that the steady equatorial current structure not only in the Pacific Ocean but also in the Atlantic and to a lesser extent (because of the geography as well as the monsoon climate) the Indian Ocean is due to this combination of thermal and wind forcing. Some deductions on the motion greater than 10° either side of the equator can be made by looking at straightforward geostrophic balance, because it is possible through varying the density to retain this balance while incorporating temperature changes. However, nearer the equator, there is a clever modelling ploy that helps to simplify models. Near the equator, the vertical component of the Earth's angular velocity, commonly given the symbol f by oceanographers, can be put equal to βy, where y is the distance in metres north *or south* of the equator (y is negative south of the equator, of course) and β is a constant dependent only on the value of the Earth's angular velocity and the radius of the Earth. This enables modellers to give a simple representation to the reversal of sign of f and to predict some of the features seen in Figure 5.3. Fortunately, in this already complicated picture, salinity gradients, although present, do not have any significant influence on large-scale equatorial current structure. Many more details of equatorial modelling including transient phenomena can be found in the excellent book *El Niño, La Niña, and the Southern Oscillation* by George Philander (1990).

5.5 Modelling large-scale phenomena

In order to study correctly any particular phenomenon, the modeller has to know what effects to retain and what effects to leave out. Let us look at a few. In order to study the details of the Gulf Stream, since the current is quite intense, and its horizontal extent is limited, the Rossby number (see Chapter 2) is not small enough for the neglect of the rather awkward advective acceleration terms. These terms are distinctive since they are non-linear and present a challenge to the modeller which is absent in other terms in the dynamic balance of the ocean. It is terms like these which cause interference between dynamic quantities of varying length scale and can result in what have been termed energy cascades from one length scale to a smaller one, and also to larger ones. The fashionable term chaos can be used to describe results from the former cascade. The latter cascade must be at least as common.

Another example is the previously mentioned Somali current, which is the western boundary current of the northern Indian Ocean. This current is very seasonal, with a periodicity linked with the onset of the monsoon. Obviously, in order to describe the dynamics of the Somali current, account has to be taken of this time variation. Retention of all the acceleration terms in the momentum balance is thus essential. It turns out that the response of the Somali current to monsoon winds can be modelled successfully merely by including these acceleration terms in the usual Coriolis versus lateral eddy viscosity relation. Of course, the variation of the vertical component of Coriolis acceleration with latitude remains vital to include since the Somali current is a western boundary current. Lighthill (1968) gives a mathematical and physical account of this modelling.

Another interesting current system already alluded to is what happens at the equator. Only in the 1960s were there models which adequately described equatorial currents. At the equator itself, there is no influence of Coriolis acceleration and either side of the equator its influence reverses in sign (see the last section). To the layman in the north, it therefore seems as if 'everything reverses in direction' south of the equator, western boundary currents travel south instead of north, etc. However, as already mentioned, in order to successfully model equatorial currents it is only necessary to include a form of the Coriolis parameter that carries the property of differing in sign either side, and zero on the equator itself. Otherwise, models of equatorial dynamics possess the same dynamic balance (Coriolis acceleration versus lateral eddy viscosity) as the rest of the ocean. What is true, however, is that there is a great deal of finer structure at the equator in terms of countercurrents and variations with depth. The equatorial undercurrent for which there is little evidence at the surface, is massive at depth. The structure of the currents in the vertical are very important and in modelling them density changes need to be taken into account. This undercurrent, a particularly fierce subsea equatorial countercurrent just south of the equator in the Pacific (running west to east), is now known to be responsible for the suppression of the economically vital coastal upwelling that usually occurs off the Peruvian coast (this suppression is termed *El Niño*; see Section 5.7.2). Understanding its dynamics via modelling thus assumes new dimensions in importance. In this rather more complicated modelling, in addition to the Coriolis terms and lateral friction, it is necessary to include the Rossby radius of deformation based on a local *equatorial* length scale. This length scale itself has within it a Brunt–Väisälä frequency arising from changes in the density of the ocean with depth. The non-modeller can legitimately question the validity of an unseemly mix of oceanic and smaller length scales in (ostensibly) the same model. Justification lies in the fact that, as is typical, it is possible to separate horizontal and vertical phenomena such that the horizontal dynamics occur on scales similar to the entirely depth-independent Somali current response, whereas the vertical scale depends on vertical density differences. In fact the horizontal structure is on a 10^5 m scale, whereas the vertical structure is on a 10^2 m scale. Mathematicians use what is called modal separation to carry out the details. The cleverness of these equatorial models lies in the dependence of the local equatorial length scale on both the Brunt–Väisälä frequency *and* the variation of Coriolis parameter with latitude, yet simultaneously the elegant mathematical separation of horizontal and vertical motion is apparent on examining the details. We will go into more technical detail in Section 5.7.

5.6 Modern ocean modelling

In the 1970s, the development of large-scale cooperative exercises revolutionised the observational side of oceanic marine science. In the last half of the 1980s, a similar revolution happened in modelling. There are several large-scale modelling exercises still very much active in the 1990s. Here large scale takes on both meanings, covering a large area of the world's oceans, as well as involving a large number of research scientists from a variety of countries. The World Ocean Circulation Experiment (WOCE) is particularly all-embracing, and many of the latest advances have taken place under this umbrella. Some subsets of WOCE, for example, TOGA (Tropical Ocean Global Atmosphere) have run their course, whereas others such as FRAM (Fine Resolution Antarctic Model) remain active.

It is worth dwelling on FRAM, which at the moment is enjoying a large modelling effort. The latest models of the Antarctic are in turn embedded in the latest global ocean models, which have a resolution of $\frac{1}{4}°$ and cover almost the entire globe. Using the new design of parallel computing, finer resolution is on its way. The all-important eddy-resolving ability of models is what is being striven for here – although old rhymes about little fleas having smaller fleas springs to mind. FRAM itself is a model that covers one-third of the entire world ocean centred on the Antarctic circumpolar current. Attention has been focused on the structure of the current, in particular its vertical temperature profiles and the distribution of stress. Other important areas of research under FRAM are the interaction of the Antarctic circumpolar current with the Agulhas current that rounds the Cape of Good Hope, the transportation of heat via meridional eddies and the steering of the Antarctic circumpolar current by the rugged topography found in that part of the ocean. This kind of large-area specific modelling places heavy reliance on fine-resolution finite difference modelling, and does not use many simplifications in terms of neglecting various effects since there is no need, the huge computers can cope, so why leave anything out? The perennial danger is that the modelling complexity obscures fundamental errors in the physics contained in the models. The underhand side of this was mentioned in Chapter 3, but here the scene is different. What we have are pure scientists who are trying to understand what are particularly complicated interrelated series of physical processes (no mention of chemistry or biology – yet!). Therefore, the use of large computer models is simply the best that one can do. It is recognised that the next major advances must come through an improved understanding of the physics of the oceanic processes; the present 'hit list' contains the ocean surface, ice, deep convection and overflows, vertical mixing and the parametrisation of eddies.

Other modelling areas of interest include our old friend the Gulf Stream, in particular why models have been so singularly unsuccessful at predicting the latitude of its separation at Cape Hatteras. A solution to this problem may also resolve why models also fail at predicting the separation latitude of the Kuroshio in the Pacific. The predictions have the Kuroshio detaching further north than it should. The time-dependent flow through Indonesia, upwelling off Arabia and a strong eddy found off Somalia are all being investigated by modellers under the continuing WOCE initiative.

5.7 Two detailed modelling examples

In this last section of this chapter, we look at a specific model in some detail. Some familiarity with numerical modelling is definitely an advantage here, but an overall picture can certainly be gleaned without this, even though some of the details may elude the reader.

5.7.1 *The Semtner model*

Since the late 1960s, much of the modelling effort in the world has been used in large international exercises. These exercises began as observational studies such as GARP (Global Atmospheric Research Programme), which included a tropical sea study, and MODE (Mid-Ocean Dynamics Experiment), which was largely US-funded and concentrated on the Gulf Stream conveniently near Woods Hole Oceanographic Institute in Boston, Massachusetts. The latest continuing programme mentioned in some detail in Section 5.6 is WOCE (World Ocean Circulation Experiment) and its large modelling offspring FRAM (Fine Resolution Antarctic Model). It is noticeable that the acronyms proliferate, reflecting an unfortunate late-twentieth-century disease. Also, modelling is playing an increasingly important role in large international exercises. Let us now introduce our example.

In Chapter 3 it was noted that, in order to model the variation of quantities with the vertical, the z variable could either be replaced with σ coordinates whereby $\sigma = 0$ was the sea surface and $\sigma = -1$ was the sea bed, or be discretised in the same way as x or y. The preferred option is to use σ coordinates, and then to adopt a modal model. The zeroth mode is all variables do not depend on depth; the first mode is a two-layer ocean, and so on. The more modes that are adopted, the more vertical the structure that can be incorporated (see Section 3.4 for more details).

The World Ocean Circulation Experiment is a worldwide exercise committed to improving our understanding of ocean physics. The goal is to be able to predict major trends and the consequences of the action of nature (e.g. large storms, earthquakes) and the action of man (e.g. global warming through enhanced emissions of greenhouse gases). Within WOCE, which by its very nature is an extremely large exercise, there is a strong modelling theme; in fact, many modelling themes. Many site-specific numerical models of great complexity are being produced. We will give some of the basic details of the models later, but it is useful here to write down some of the known characteristics of particular parts of the deep ocean that site-specific models will have to accurately describe. It has long been widely recognised that the Southern Ocean plays an important role in overall ocean dynamics. It has been known for a long time that nutrient-rich Antarctic bottom water finds its way far north, sometimes even crossing the equator, and revitalises the biology of the Indian, Atlantic and Pacific Oceans, although details of precisely how this happens are lacking. There is much less in the way of land mass south of the equator; in particular, there is a zonal land free passage through which flows the Antarctic circumpolar current. Because this current is zonally unbounded, it transports large amounts of heat and salt, but again we do not know precisely how. There are also other important Antarctic water masses, each with its own characteristics and each with an important role to play in the general circulation of the ocean. The FRAM exercise is a part of WOCE that is virtually all modelling. Its brief includes assessing the role of bottom topography in controlling the flow of the above-mentioned Antarctic circumpolar current, the modelling of net transport of water between the Atlantic, Indian and Pacific Oceans by the Antarctic circumpolar current, the polewards transport of heat, and the generation of new water masses. These are such fundamental questions that one could be forgiven for thinking that very little work had been done on modelling ocean currents of this size. That this view is wide of the mark will be obvious by a close examination of one of the early numerical ocean models. Anyone interested in more details about WOCE should write to the Southampton Oceanography Centre. This information is given hesitatingly since it may be inappropriate for international readers. Anyway, the WOCE office issues a (free) international newsletter that keeps interested scientists abreast of the latest information in both modelling and observations on this large scale.

We shall focus on the model developed by Bert Semtner in the mid-1970s. It is essentially a two-dimensional model, but that does not mean it is all that simple! It is based on spherical coordinates, latitude and longitude, and the equations are integrated through the vertical from the sea bed ($z = -H$) to the sea surface ($z = 0$). All non-linear and turbulence terms are included in the model, although the age of the model becomes very apparent when one sees that the turbulence is modelled via a constant eddy viscosity. This of course prompts the standard query over what value to use. Unrealistic closed north and south boundaries are used, and the east and west boundaries were either closed (also unrealistic) or periodic (which is fine for the Antarctic circumpolar current). The sea surface boundary conditions were quite realistic inasmuch as they included salt and temperature (at least in theory) and could even consider evaporation and precipitation. It would be inappropriate in a non-specialist text such as this to write down the equations in their full glory, even given the stated simplifications. Instead, we shall look at a few terms in the horizontal momentum balance that is valid in the east–west direction. The hydrostatic balance is still valid and takes the form

$$\frac{\partial p}{\partial z} = -\rho g,$$

i.e. vertical pressure gradient, the left-hand side, is density (ρ) multiplied by acceleration due to gravity (g). The negative sign arises because z is positive upwards, whereas pressure increases downwards. This enables the pressure terms in the equation of motion (to return to the east–west equation mentioned above) to be eliminated through the device of differentiating with respect to z and substituting $(-\rho g)$ whenever $\partial p/\partial z$ occurs. Once this has been done and all pressure terms have gone, the equations are vertically integrated so that, for example, u, the easterly flow, is replaced by $\bar{u}$, where

$$\bar{u} = \frac{1}{H} \int_{-h}^{0} u \, dz.$$

$\bar{u}$ is then only a function of latitude and longitude. A similar procedure defines $\bar{v}$, the northwards vertically integrated flow. Two bonuses occur as a result of doing this vertical integration. First, both $\bar{u}$ and $\bar{v}$ can be expressed in terms of *one* unknown function (called the transport stream function, but never mind about the name). Also, there are terms in the equations that contain the vertical integration of turbulent (Reynolds) stresses. At the surface, $z = 0$, and at the sea bed, $z = -H$, there are boundary conditions that naturally contain the Reynolds stress and how it relates to gradients of the flow. These boundary conditions thus become explicit terms in the governing equations (for the above-mentioned stream function). Finally, salt and temperature are similarly treated and this time, boundary conditions that may contain information on evaporation and precipitation also become explicit terms in the equations of motion. Once these simplifications have been achieved, finite differences are applied and the differential equations, now transformed into difference equations (which are algebraic simultaneous equations) are solved using the Arakawa B grid (see Section 3.4.3, Figure 3.5). The early Semtner models concentrated on trying to resolve eddy motion in the 'open' ocean. We have made no mention here of the strenuous efforts that Semtner takes to make sure that energy is conserved. This remains one of the most awkward problems for ocean modellers. The boundary conditions control the input and output of energy. Friction is an energy sink at the sea bed and at the coasts, whereas heat and wind can input energy at the sea surface.

Other models have used Semtner's model as a starting point and have addressed some of the simplifications he made, in particular the treatment of boundaries and, at least for limited area models, the parametrisation of turbulence. These more

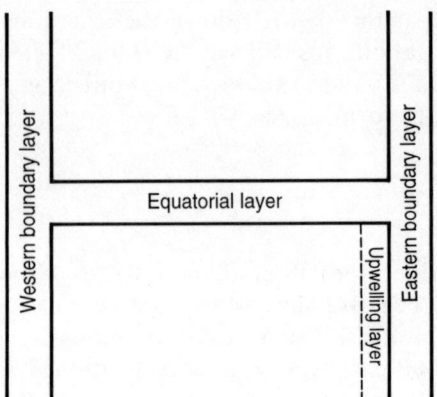

Figure 5.4 A simple picture of equatorial boundary layers.

sophisticated models tend to be site specific, which takes us right back to WOCE and FRAM, where this section started.

5.7.2 *El Niño*

El Niño (Spanish for 'the Christ child') is the name given to a warm current that appears each December in the eastern Pacific Ocean, and flows southwards along the coasts of Equador and Peru. Nowadays the name has come to be used only when this current is so strong that it overcomes the normal flow pattern, causing catastrophic fatalities in the fish (*anchoveta*) population and bringing economic disaster to Equador and Peru. These *El Niño* events tend to occur every three to seven years, and it is obviously in everyone's best interests to understand what triggers them. Modelling *El Niño* has been underway for about 25 years, and most oceanographers agree that there is now quite a good understanding of the overall dynamics. Prediction of *El Niño* is not yet, however, one hundred percent certain.

In order to model the onset of *El Niño* reasonably successfully, there are various ingredients that need to be present. First, it is now accepted that correct dynamic modelling of the equatorial undercurrent is essential. The equatorial undercurrent is that part of the Pacific equatorial current system that is sub-sea south of the equator and flows from west to east (see Section 5.4). The dynamics of the equatorial circulation can only be faithfully reproduced in a model that takes into account the change in the vertical component of Coriolis acceleration with latitude, the so-called β-effect. The term β arises from considering the Coriolis parameter f, defined as

$$f = 2\Omega \sin \theta,$$

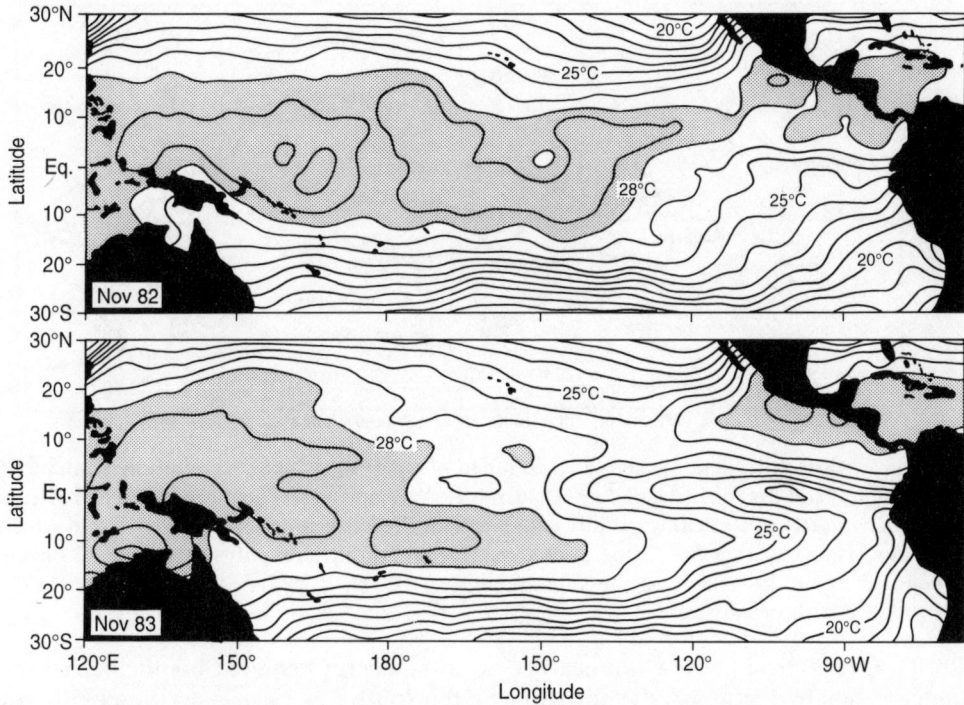

Figure 5.5 Sea surface temperatures in November 1982 during *El Niño*, and November 1983 during *El Niña*. From Rassmussen and Carpenter (1982). Reproduced with permission.

where Ω is the angular velocity of the Earth and θ is the latitude, and approximating its variation close to the equator (where θ is small) by

$$f = \beta y,$$

where y is the northwards distance from the equator, and $\beta = 2\Omega/R$, R being the radius of the Earth. If the equatorial layer is incorporated into a model which also includes the western boundary layer, with its intense current, and an eastern upwelling boundary layer, then an overall dynamic picture emerges, as shown in Figure 5.4.

The equatorial layer contains the important equatorial undercurrent as well as other east to west equatorial currents, although these latter tend to be more surface currents and not so important in the dynamics of upwelling. For it is the suppression of the upwelling at the eastern side of the ocean that is primarily responsible for the devastating effects of *El Niño*. The obvious question to ask therefore is why the equatorial undercurrent is so strong in some years so as to completely obliterate east coast upwelling, yet in other years the upwelling occurs without any hindrance from it. Modern models of the Pacific Ocean equatorial region include the sea surface temperature, and it is thought that a particularly warm sea surface in the area of the

58   Modelling Marine Processes

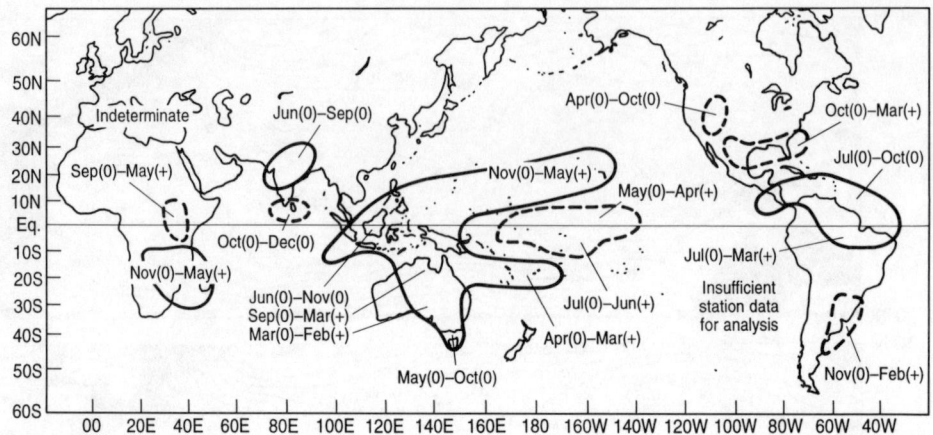

Figure 5.6 Schematic diagram showing the different regions in which precipitation is enhanced (dashed lines) and diminished (solid lines) during *El Niño* episodes. The months, which indicate when the regions are affected, generally coincide with the local rainy season. The year in which anomalously high sea surface temperatures first appear and then amplify in the tropical Pacific is denoted by (0); (+) refers to the subsequent year. From Ropelowski and Halpert (1987). Reproduced with permission.

Pacific Ocean around New Guinea provides the energy required for the equatorial undercurrent to overpower the upwelling off the South American coast. Some support for this is shown if the observations of sea surface temperature in the tropical Pacific for *El Niño* years are compared with those over the same area in an opposing *La Niña* year (see Figure 5.5).

The term *La Niña* refers to a situation in the Pacific ocean where the sea surface temperatures in the central and eastern tropics are unusually low and the trade winds unusually intense. It is the oceanographic opposite to *El Niño* (the (male) child) and means the (female) child. It does not have a similar impact on the climate or economy as *El Niño*, but it is as dynamically significant. It is the opposite phase of a process that has been termed the southern oscillation. The acronym ENSO has been used for *El Niño*–southern oscillation, but let us not succumb to more acronyms! It is therefore as important to understand *La Niña* as it is to understand *El Niño*, and to build a model of the southern oscillation in its entirety, if possible. The book *El Niño, La Niña, and the Southern Oscillation* by George Philander (1990) contains some of the latest ideas, but this is still a very active research area. In fact the southern oscillation has a global impact, as can be seen in Figure 5.6. This figure shows regions of enhanced and diminished precipitation alongside the months of the year in which they occur. In general, the enhanced precipitation coincides with the local rainy season, but during *El Niño* episodes, which occur at times coincident with the records shown dotted, the higher sea surface temperature coincides with enhanced rainfall.

To return to more general modelling, there is no doubt that the principal driving force behind the equatorial currents and therefore the dynamics of *El Niño* is the wind stress, and it is this input that remains the main surface boundary condition for numerical models. The temperature field is, however, a necessary input to models if these models are to be used to predict the nature of the southern oscillation and therefore *El Niño* events. It is tempting to include other world climate catastrophes under the umbrella of *El Niño*. For example, in 1982–84, which contained the *El Niño* year 1982 (a particularly dramatic one), there were the following extreme weather conditions worldwide: unusually warm summers in Europe, northern Russia, and the USA, and drought in Australia, Africa and parts of North and South America. In 1992, another strong *El Niño* year, there were floods in Texas accompanying an unusually mild winter. As a modeller it is entirely logical to think of the world as a single entity which should be modelled as a whole; the modellers' version of the 'Gaia hypothesis', perhaps. One therefore has to be careful in drawing too many direct links between climate extremes that happen to occur simultaneously. A much safer procedure, and one more in tune with the scientific method, is to examine the occurrences of *El Niño* (and its opposite weather pattern, *La Niña*) as part of the southern oscillation. This model of the southern oscillation is then taken together with the occurrence of any chosen climate extreme (hot spells or particularly wet periods) and their concurrence tested statistically. At the moment, not enough data are available for any significant conclusion to be reached, so the modeller must build dynamically accurate models of each phenomenon until the ultimate all-inclusive model comes along. The date for that to happen seems as far away as ever! The mathematical details of the modelling of *El Niño* and the southern oscillation have purposely been omitted as they are not easy to understand. More details are given in George Philander's book, and in Appendix A.

Chapter Six

Continental shelf sea modelling

6.1 Introduction

Oceans have a general structure which is dictated by their geological origin. There are mid-oceanic ridges, trenches and, most important for us here, continental shelves. These shelves border the margins of the oceans, the seas over them are only about 200 m deep, and the transition from this 200 m to the oceanic 3000 m is achieved through a relatively small region, the continental slope – which is aptly named, since the slope is commonly only 4°. Simple trigonometry shows that the depth can sink from 200 to 3000 m in a horizontal distance of 40 km with a slope of 4°. Since the width of the ocean is measured in thousands of kilometres, and most continental shelves are hundreds of kilometres wide, this slope region is of insignificant horizontal extent. It is, of course, only the dimension that is insignificant. The continental slope contains important currents and is a significant source and sink of energy for many different types of flow. However, for continental shelf modellers it serves only to mark the border between deep ocean and continental shelf, and is the site of the open boundary condition. The dimensions of the continental shelf render models of continental shelf seas quite distinct from ocean models.

One of the most distinctive features of continental shelf seas is the relative strength of the tidal currents compared with those that arise from other sources. From the point of view of fluid flow this feature is easy to explain. Just as flow through a large-diameter pipe accelerates to a faster value if the diameter decreases, so a slow tidal flow in the deep sea accelerates to fast tide over the shallow continental shelf. In shelf regions, tidal currents are usually ten times stronger than currents from other sources (wind or convection due to freshwater inputs from rivers). Let us begin this look at continental shelf sea modelling with tidal models.

6.2 Tidal modelling

At the outset, let us declare that we shall be concerned with practical tidal dynamics and not with the kind of theoretical tidal modelling which was published a hundred

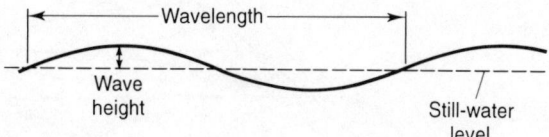

Figure 6.1 A surface water wave.

or more years ago, dating back to Laplace in 1776, involving periodic hydrodynamics on a rotating sphere.

Tides are due to the astronomical forces that arise from gravitational attraction between the Earth and the Moon and, to a lesser extent, the Earth and the Sun. These forces themselves act on the Earth's large bodies of water, the Pacific, Atlantic and Indian Oceans. Typically, the Moon pulls the surface of the Atlantic Ocean about half a metre from its mean level both when it is directly overhead, and, paradoxically, when it is on the other side of the Earth too. This slight tide, and its equivalent depression when the Moon is on either horizon, causes a wave (two high waters a day, as the Earth rotates) which propagates. This wave passes over the continental shelf and, once on the continental shelf itself, increases in amplitude because of the conservation of mass (the same reason as the enhancement of tidal velocities). Another feature of tides is their wavelength. This is the horizontal distance between two successive high waters at any fixed time. This distance is typically 1000 km (for a tide with two high waters a day in a continental shelf region). The amplitude of a tide is but a few metres or even less. The wave slope, a common dimensionless parameter which is often used as a characteristic measure, is the ratio of these two quantities (see Figure 6.1) and is thus only about 10^{-5}. This means that tides, considered as water waves, are very long waves indeed. Some elements of shallow water wave theory were given in Chapter 2. There, non-dimensional quantities were defined, and as mentioned, these apply very well to tides. In order to see how shallow these water waves are let us insert some typical values for the parameters. Since the depth of the continental shelf sea is only 200 m or so, and the ratio depth/wavelength is only 2×10^{-4}, we can consider the water as indeed very shallow. One consequence of this is the tidal currents should be virtually independent of the depth. This is certainly true if we are concerned with tidal height predictions. However, tidal currents are influenced by bottom friction due to the roughness of the sea bed; this can be of crucial importance, and we will return to it later.

The tide can therefore be crudely modelled as a to and fro, depth-independent movement of the sea. Since the pressure is hydrostatic, that is, only dependent on the weight of water above a particular point, the pressure is proportional to the elevation of the sea surface above mean sea level (taking the mean sea level as a reference level for pressure). Hence horizontal pressure gradients are proportional to the slope of the sea surface. The dynamic balance required to model tidal motion in shallow water (continental shelf seas) is therefore a match between particle acceleration

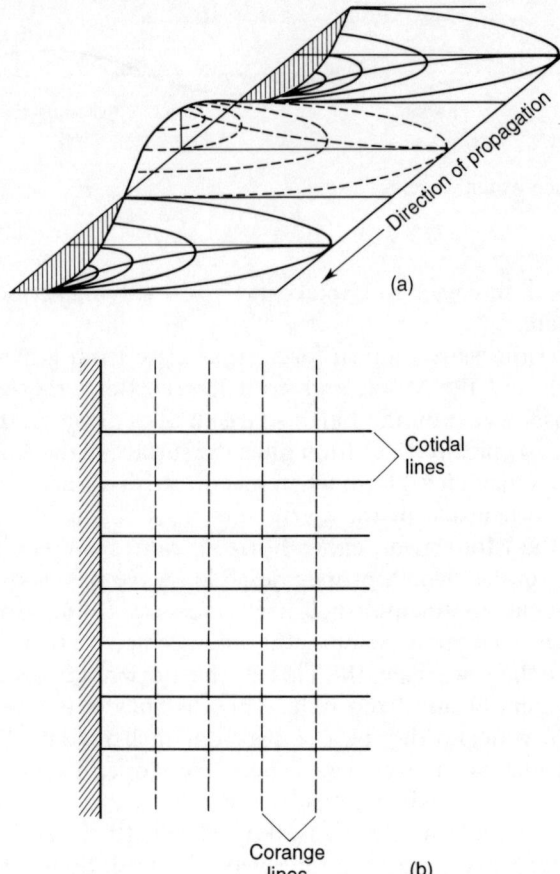

Figure 6.2 A Kelvin wave: (a) isometric view; (b) plan view.

due to the back and forth water movement of the tide together with Coriolis acceleration and the pressure gradient forces which manifest themselves as a sea surface slope. In a narrow inlet or estuary, tidal waves propagate up or down in more or less the same way as waves in a domestic bath can be driven by a convenient wiggle of a foot. However, tidal waves have a very long wavelength. This is as if the water in the bath rises and falls, keeping the surface horizontal (conservation of mass demands an infinitely long bath here!). All frictional effects have been ignored of course. The introduction of the Earth's rotation in the form of the Coriolis acceleration causes this wave to behave rather differently. It is as if the wave in the bath, instead of preserving a horizontal crest as high water comes toward you, is lopsided to the extent that the crest is higher to the left, falling away exponentially to the right. As it travels away from you, the highest point of the wave is on the right and the line of high water falls away exponentially to the left. Viewed from above the wave hugs the edge of the bath, propagating anticlockwise around it. This direction is reversed in the southern hemisphere. Of course, since the bath has a finite width and length, neither the exponential profile offshore nor the sinusoidal nature of the wave along the shore is perfect. Perfection only occurs with an infinitely long coastline next to an infinitely wide sea. This perfection is called a Kelvin wave, and although it is never found in pure form, the tides in many coastal seas, e.g. the northern North Sea, are well approximated by them. Figure 6.2 shows a Kelvin wave together with a plan view in terms of cotidal lines and corange lines. The cotidal line is a line along which the maximum tidal amplitude is attained at the same time, and it can be thought of as an instantaneous snapshot of high water taken at 3 hourly intervals. The corange lines are lines along which the range of the tide is the same. For a Kelvin wave, these lines form a square grid. Let us see how this square grid arises. The wave travels down the coast with its crest perpendicular to the coastline; hence cotidal lines, which must follow the crest line, are also perpendicular to the coast. On the other hand, along a line parallel to the coast the 'tide' (that is, how much the sea surface rises and falls) will be the same. This is precisely the range, and hence corange lines are also parallel to the coast. Now compare this idealised square grid system of cotidal and corange lines with an actual picture of cotidal and corange lines taken from observations in the North Sea (Figure 6.3). You will see that, near the coast of the UK, in the north, the cotidal and corange lines actually form a square pattern. This tide is almost a perfect Kelvin wave, although perfection is absent elsewhere.

In order to predict tidal elevations and the currents that arise from tides, it is usual to build a numerical model based on a grid designed to cover the desired basin, sea or coastal area. Before discussing this side of modelling, let us look at some of the processes it is essential to incorporate in a successful tidal model. Perhaps the best place to start is to see what can be ignored. First, it can be assumed that there is no weather! This rather startling assumption is put in context once it is realised that wind and pressure effects that arise from the weather can be added later. They are not essential ingredients to a tidal model, although everyone is aware of their importance to understanding flooding due to storm surges (see Section 6.3). The technical reason why we do this adding of effects is linked with the extremely long wavelength of tides (Figure 6.1). Tides are linear to the extent that meteorological effects can be added and there is very little dynamic interaction between wind and tide. Tides themselves, however, can exhibit highly non-linear characteristics. One only needs to think of shallow water where drying can occur, or peninsulas where a tidal current can change sharply in direction. Both of these scenarios have length scales within them that render the advective term (see Chapter 2) large enough to be important. When this term is important, the model is described as non-linear. However, in the modelling of tides there is no non-linearity connected with wind or pressure. (Contrast this with surface sea waves and wind which can and do interact and give rise to the phenomenon of Langmuir circulations; see Section 6.4.) Second, unless internal tides are of concern, it is reasonably safe to ignore changes in density. In coastal seas, density changes are normally due to freshwater input via rivers rather

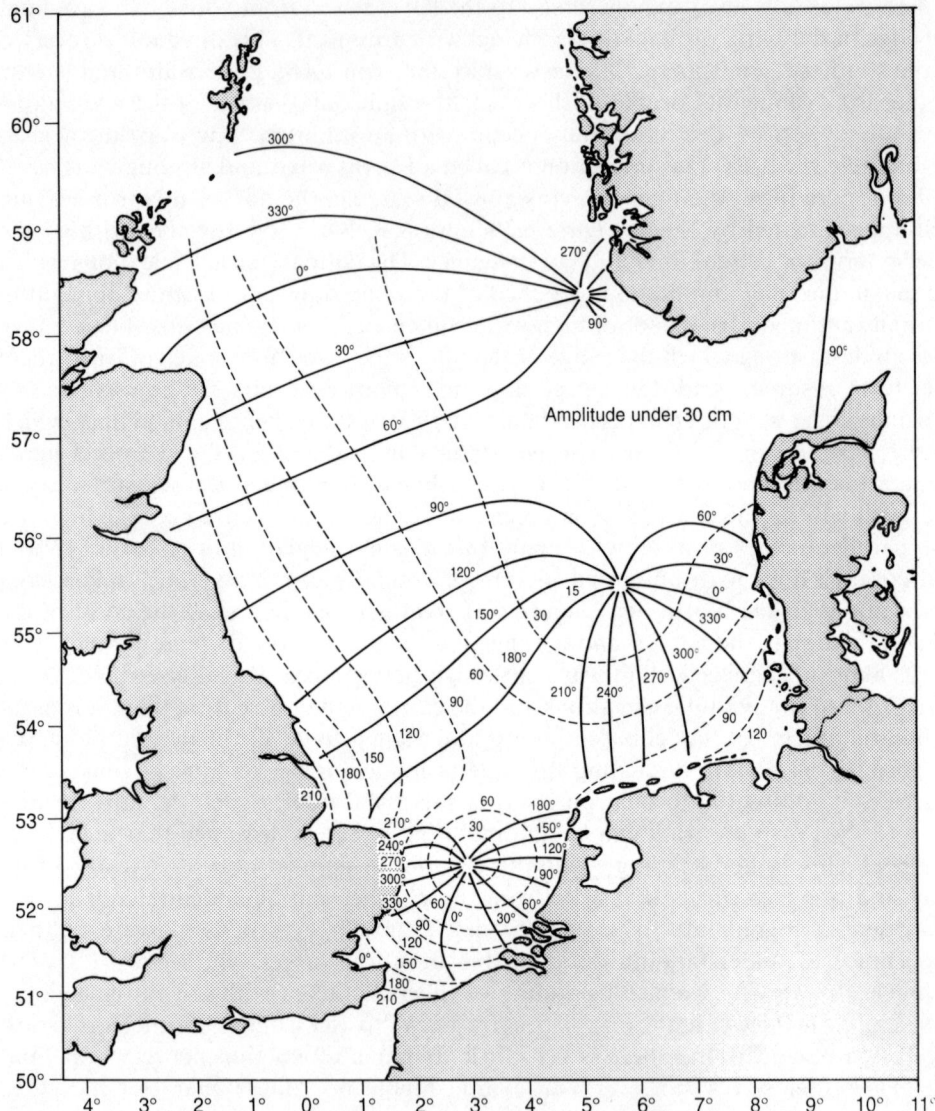

Figure 6.3 Cotidal and corange lines, denoted by full and dotted lines, respectively. The associated numbers give the values of γ (in degrees) and of H (in cm).

than to temperature contrasts, although summer thermoclines do exist. Again preliminary tidal modelling can be done accurately without involving these density fronts or pycnoclines. This is due principally to the difference in length scales associated with these density interfaces. These issues will also be addressed in Section 6.4.

If the prime concern is modelling tidal elevation, certainly the most important factor for shipping, then the model must include the following factors: particle acceleration (because the tide is a wave, albeit a very long one, and waves have a to and fro motion which implies acceleration); Coriolis acceleration (because the length scale of the waves is in general large enough for the rotation of the Earth to be important); sea surface slope (this slope is alone responsible for the horizontal pressure gradients that act as a horizontal force); and friction. This last term needs some explanation. Astronomical forces generate the tides, but as far as the seas of the continental shelf are concerned, they are driven by the oscillations of a larger neighbouring body of water (the ocean). The only mechanism available to dissipate the energy that is being so liberally pumped in via the continental slope is friction, and this frictional dissipation has to take place at the sea bed. The form that friction takes has been outlined in Chapter 2. Normally, a quadratic law relating the drag to the square of the speed provides an adequate model that can also be justified on dimensional grounds.

6.3 Wind-driven circulation

The other main cause of continental shelf sea currents is wind. The balance valid near the sea surface is one between the Coriolis acceleration and friction which causes Ekman circulation. These spiral currents are not easy to detect by measurements, since their structure is easily disturbed by local effects such as eddies and surface waves. The major influence of the wind on the ocean occurs, quite obviously, close to the surface of the sea. However, the dynamics of the ocean are complex and the fact that the surface of the sea is moving under the influence of the wind in turn drives deeper flows. In particular, there is a vertical velocity which results directly from the wind, and this vertical velocity when occurring at the base of the Ekman layer is termed *Ekman suction*. A positive or negative Ekman suction will exert a strong effect on that part of the sea that is far below the surface and is not directly wind driven.

The Earth rotates, and a consequence of this rotation is that wind-driven flows do not move in the direction of the wind. In the Ekman layer, the net effect of a steady wind is a flow to the right of the wind (in the northern hemisphere; left in the southern hemisphere). Wind is never steady for long periods, so the Ekman currents are always being spun up or attenuating. Since the detailed structure of a pure Ekman current is spiral in nature (see Figure 6.4), even given the mythical steady wind, the real waxing and waning versions can bear a close resemblance to turbulent eddies.

So far there has been no mention at all of coasts. Coastlines are a barrier to currents and all motion is forced to flow parallel to them. Modelling currents along coasts tends to be site specific; there is very little to be said of a general nature, except that again the overall dynamic balance remains between frictional forces, Coriolis acceleration and pressure gradient forces. When there are bends in the coastline on a scale of hundreds of kilometres, then it is usually prudent to consider additionally

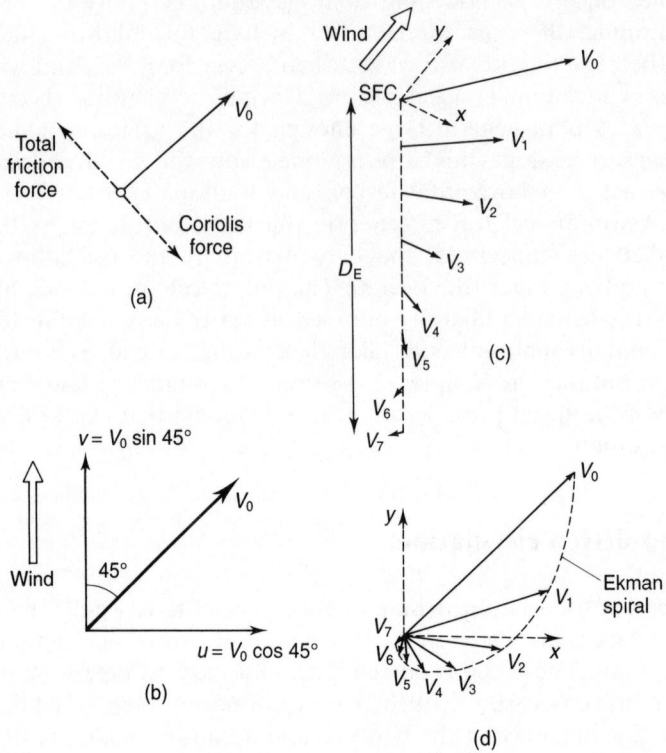

Figure 6.4 Structure of a pure Ekman current. (a) Forces and surface velocity; (b) plan view at surface; (c) perspective view; and (d) plan view, velocities at equal depth intervals, as in (c). From Pond and Pickard (1991). Reproduced with permission.

the effects of advective acceleration. An effective way of assessing whether or not various terms should or should not be included in a particular model of a specific sea is to utilise dimensional analysis, as outlined in Chapter 2.

Another important feature of continental shelf seas are river mouths. The circulation in the saline mouths of many river outlets resembles that of a coastal sea. The Moray Firth and Firth of Forth in Scotland, and the mouths of the larger Norwegian fjords are good examples. Further upstream, the presence of fresh water exerts a greater influence on the circulation.

6.4 Density-driven and wave-driven flows

Salt water is heavier than fresh water. This gives rise to the 'saline wedge' picture often taken as typical of an estuary (see Figure 6.5). Modelling estuarial flows has a long tradition. The first approach, still of some use today, is to use box models. In

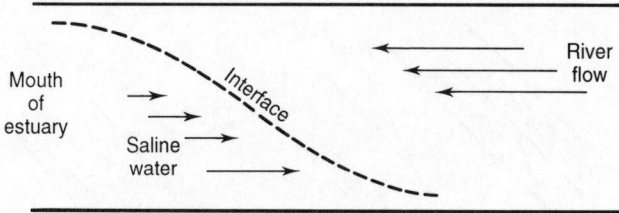

Figure 6.5 Saline wedge in a stratified estuary.

these models, the domain of the estuary is divided into conveniently sized boxes (either two-dimensional vertical or horizontal sections or three-dimensional boxes), and an estimation is made of the transportation of various water properties across the boundaries of each box. These properties can include salinity, temperature, various chemical and biological constituents as well as flow. Once the boxes have their assigned parameter values, and each box contains water with properties that are consistent with the properties of the water in adjacent boxes and consistent with an overall agreed picture, then the model can be exploited. For example, an inflow at one point of the estuary will influence the water properties in one box or two, but will also probably influence the behaviour of the water in surrounding boxes through having to satisfy the conditions at the boundary of the boxes. These box models are still useful under two circumstances: first, if an approximate picture is required quickly, then a box model can give some insights; and second, if there is no appropriate software available to build a more elaborate model and, more importantly, only a sketchy knowledge in terms of observations in certain areas, then a box model can provide a first rough model. Box models can often be the initial models that trigger additional observation programmes or experiments. In an educational context too, box models can be a useful vehicle for introducing the rudiments of modelling. However, it must be said that if reliable predictions are the order of the day, then more sophisticated models which make use of software would normally be required.

Langmuir circulation is the name given to vortices that occur near the surface of the sea. A few years ago, it could be said that not everyone agreed about their existence, but it is now accepted that they are always present when the wind is stronger than 7 metres per second. What are these vortices? They are near-surface circulation that arises as a result of the interaction between the waves on the surface of the sea and wind-driven flow. In typical Langmuir circulation (see Figure 6.6) the fluid moves in a path reminiscent of a coiled spring, spiralling with the axis of the spring horizontal and more or less parallel to the wind. The vortices have a spacing of about 200 m in the sea, although analogous circulation is found in lakes and here the spacing is much less, in fact only in the order of tens of metres. Detailed modelling of Langmuir circulation is not easy since very different mechanisms need to be incorporated to include both wind-driven currents and currents due to waves. When modelling waves, the first assumption is that the profile of the waves themselves can be approximated

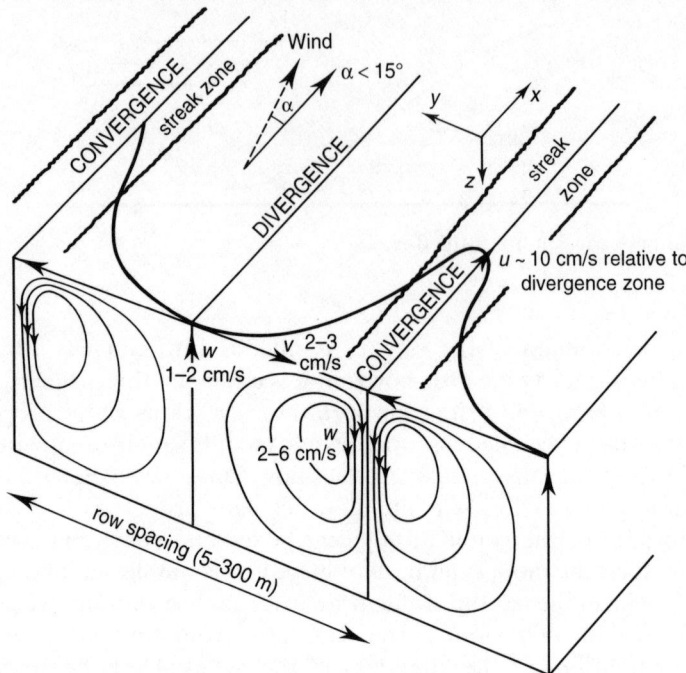

Figure 6.6 Langmuir circulation.

by a sinusoidal shape under which the water particles follow closed circular (or nearly circular) paths; see Figure 6.7. Unfortunately, such a model of waves cannot predict any net movement of water or indeed waterborne material. If this is surprising, look at a cork bobbing on a wavy sea; it may have a very small net drift, but the principal movement is circular. This is the same for all the water particles. The assumption of a purely sinusoidal surface wave in fact irradicates any wave drift – there is no wave drift under a sinusoidal sea. It is thus essential to include the non-linear or advective terms in any model that is to predict transport of material. It is this combination of a non-linear wave theory, which can predict a drift (the wave drift or Stokes' drift), with a direct wind drift that can successfully model Langmuir circulation. The dimensionless number associated with wave theories is the wave slope which is the ratio of the wave amplitude to the wave length. An appropriate dimensionless number associated with pure wind drift is the vertical Ekman number for flows whose horizontal scale is typically hundreds of metres. Such a scale is, however, usually inappropriate. The only alternative is to examine the pure wind-driven current in terms of the direct action of the air on the sea. This can be done through the stress–rate of strain relationships of the type used to define eddy viscosity (see Chapter 2). We then look at the ratio of this wind drift to the pure wave or Stokes' drift. If this

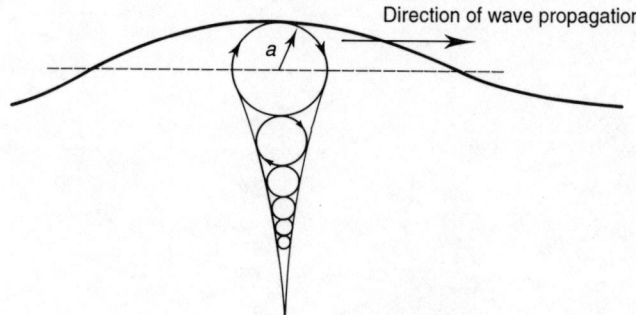

Figure 6.7 Particle motion under a deep water wave: a is the wave amplitude.

number is around unity, then we can expect both of these effects to be equally important. A model that includes all the relevant non-linear wave and wind drift terms is then built, and Langmuir circulation is predicted. Another approach is to use the Froude number, which denotes the relative importance of current (in this case wind-driven flow) and wave speed. This then leads to a stability criterion for Langmuir circulations and takes us outside the scope of this text. Perhaps the only good news is that the scale is such that the Coriolis terms can be safely ignored since the typical length scale, even of the largest Langmuir cells, is still deemed small when compared with the length scale associated with the Coriolis acceleration (this is the Rossby radius of deformation, U/f, where U is a typical speed and f is the Coriolis parameter, which is of the order of hundreds of kilometres; see Section 2.5).

The main reason for dwelling on Langmuir circulation, interesting as these three-dimensional currents are in their own right, is their importance to the ecology of the sea surface. As can be seen from the picture of Langmuir circulation (Figure 6.6), there are surface convergence zones. These zones can cause agglomerations of food as nutrient material is caught up in the circulation; at sea, lines of birds can often be seen taking advantage of these nutrient streaks. On the other hand, the same streaks are responsible for concentrating pollution such as spilt oil. Aerial photographs of oil spills also reveal these less wholesome streak lines. Hitherto non-toxic levels of pollution can be concentrated up to toxic levels by this Langmuir circulation mechanism.

Returning to consider larger seas, for example, the Irish Sea in the UK, or Chesapeake Bay in the USA, convectively induced circulation can play an important part, and Langmuir circulation plays no part. The overturning effect caused by the outflow from a large river meeting saline water can completely dominate the circulation, producing onshore flow near the sea bed and a compensating offshore flow close to the surface.

Models of coastal flows can therefore be highly complex and may include tide, wind, as well as density effects. Let us finish this chapter with some examples taken

70   *Modelling Marine Processes*

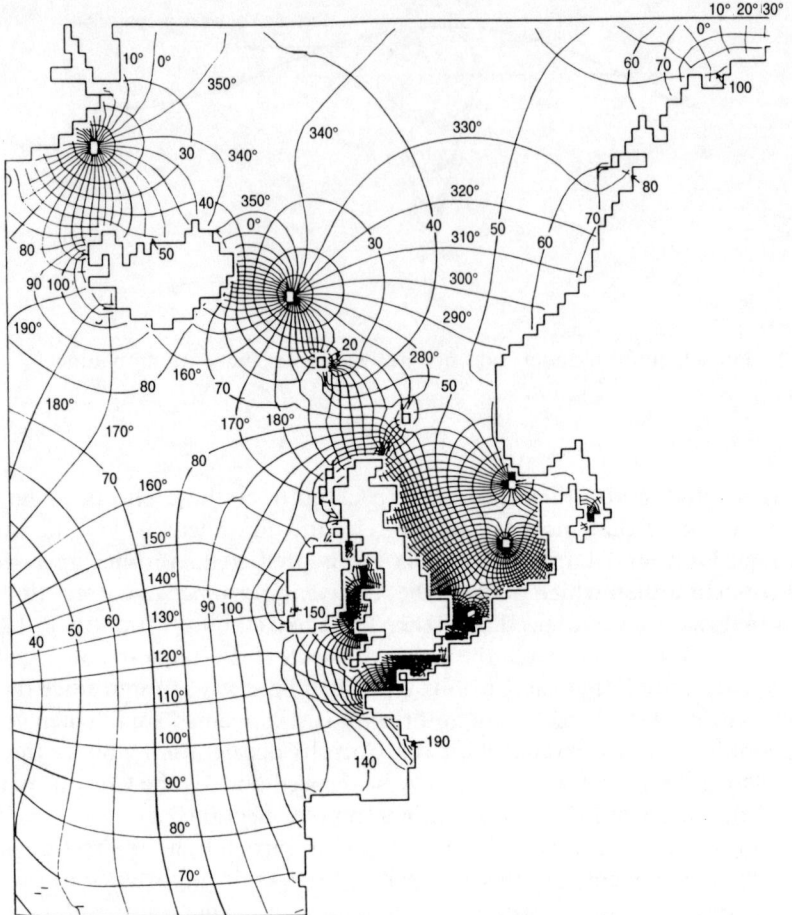

Figure 6.8   The output from a numerical tidal model.

from the research literature. This is the best way to see how to implement the modelling procedure in these quite small but highly complex areas.

6.5   Two modelling examples

Let us examine two specific examples in some detail. The first of these is the North Sea. North Sea modelling has a long and distinguished tradition dating back at least as far as the 1960s. Many papers and books have been written on how the North Sea circulation operates. Present models include wind effects and tides, all integrated and on a variety of scales (none includes Langmuir circulation, however, as this is on too small a scale).

6.5.1 The North Sea

The tides of the North Sea are largely M_2, that is, they are governed by the Moon and have two high waters a day. The accepted pattern of cotidal and corange lines is shown in Figure 6.3 and was established in the 1920s. The three points where the tide vanishes, called amphidromic points, have the cotidal lines radiating outwards and rotating clockwise. The net picture is, loosely, of a wave entering the North Sea keeping the coast on its right and exiting along the Norwegian coast. This wave is often very close to a Kelvin wave, especially along the Scottish and Northumbrian coasts (see Section 6.2), although it becomes more and more distorted as it travels round the North Sea. This distortion is caused both by irregularities in the sea bed and coast, and by frictional effects. A purely tidal numerical model does, however, predict the cotidal and corange lines of the North Sea with some accuracy (see Figure 6.8), even using a grid spacing of $\frac{1}{3}°$ latitude by $\frac{1}{2}°$ longitude. However, it is really only of theoretical interest to produce a purely tidal numerical model. It is far better to solve the equations *once*, incorporating both tidal and meteorological forcing. It is difficult to include density effects too, since these operate on a substantially different time scale (seasonal) and are very localised in their influence. One of the first three-dimensional models of the North Sea was built by Norman Heaps in the late 1960s for the specific purpose of storm surge forecasting. Storm surges are enhanced high waters caused by a combined effect of wind and tide. Only quite sophisticated models will be capable of predicting storm surges with accuracy, and these models will have to contain tidal dynamics together with wind-induced currents and the pressure field due to low-pressure centres. The tidal model that produced Figure 6.8 was accurate enough to give the amplitude of the tide within 10% and the phase to within $10°$ of observed values. In order for a model such as this to function satisfactorily, a condition must be applied at the open boundaries of the model that allows disturbances to propagate out from the model domain without anything being reflected back. Of course, any reflections would be highly unrealistic. Such a condition is called a *radiation condition* by modellers. There is a horizontal eddy viscosity imposed in such models, and a typical value is 10^3 m^2 s^{-1}. It may seem satisfactory to restrict such models to operate in only two horizontal dimensions, but this is not so. Three-dimensional models are often required since only in such models can the dissipation of momentum by friction at and very near the sea bed be correctly simulated. One common assumption is to use the quadratic friction law, as mentioned in Chapter 2.

The prime purpose for building such sophisticated tidal models is for use in modelling storm surges. Obviously, in order to predict storm surges, it is very important to combine this accurate tidal model with an equally good model that computes the elevations due to wind and air pressure. The flow chart of Figure 6.9,

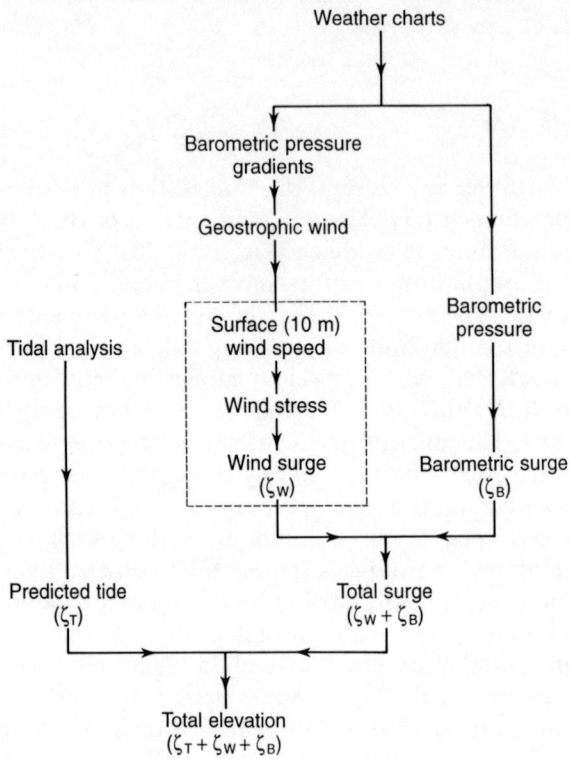

Figure 6.9 A storm surge flow chart.

due to Norman Heaps (1987), gives the procedure for the composition of a storm surge prediction model.

Over the last 25 years or so, the North Sea surge prediction models initiated by Heaps, which were adopted nationally by the UK in 1978, have been modified, updated and generally improved (principally by Roger Flather and his colleagues at the Proudman Oceanographic Laboratory). Present-day models include a representation of non-linear processes. The hydrodynamic equations are solved by using finite difference techniques. A mesh is overlaid over the North Sea (either the $\frac{1}{3}°$ latitude and $\frac{1}{2}°$ longitude grid, or a finer one) and the elevation is calculated at the centre of each grid rectangle, the velocities being evaluated either at the corners or at the mid-points of the sides. Even though non-linear advective terms are now taken into account, the equations are integrated vertically. The quadratic friction law appears as the boundary condition that arises from the vertical integration of the equations as does the wind stress at the sea surface. Figure 6.10 shows an example of how well

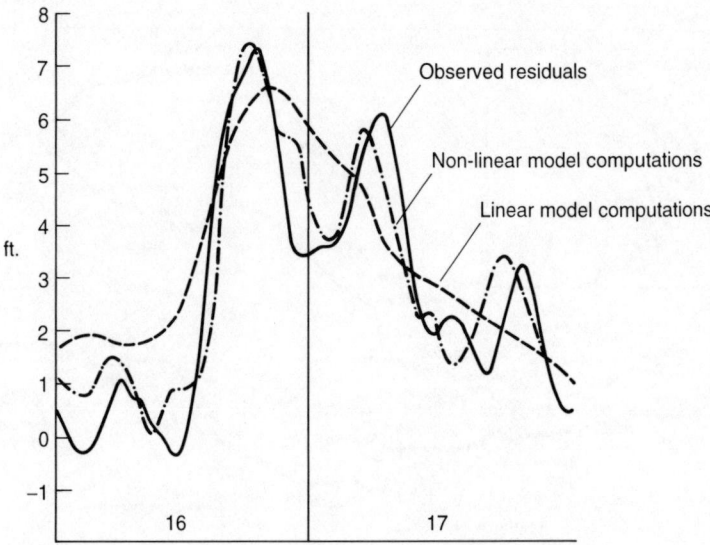

Figure 6.10 Wind-induced surge levels at Southend, 16–17 February 1962. From Banks (1974). Reproduced with permission.

such models can work (this graph dates from a 1974 model). Some simple computations are introduced in Chapter 9.

6.5.2 *The Firth of Forth and the Forth Estuary*

The main contrasts between coastal sea models and models of inlets and estuaries arise, not surprisingly, because of differences in scale. In this example, we look at the Firth of Forth on the east coast of Scotland. This particular inlet has some interesting features, perhaps the most remarkable being a natural division at Queensferry, to the east of which the Firth of Forth is virtually as saline as the North Sea and the dynamics resemble that of an embayment rather than an estuary (see Dyke, 1987). To the west, there is the Forth Estuary which has all the characteristics of a typical British estuary (see Webb and Metcalfe, 1987). Figure 6.11 shows the geography of the Firth of Forth.

Because of the geography, the models tend to be built in two parts: one to the west of the Forth road and rail bridges at Queensferry, and the second to the east. To my knowledge, there are no complete models of the Firth of Forth covering the area to the west of the bridges. Part models have been built by various engineering companies as part of managing the environment in terms of waste water and the construction of marinas for leisure, etc., although these have not been coordinated and many give contradictory findings. Others are not available in the open literature.

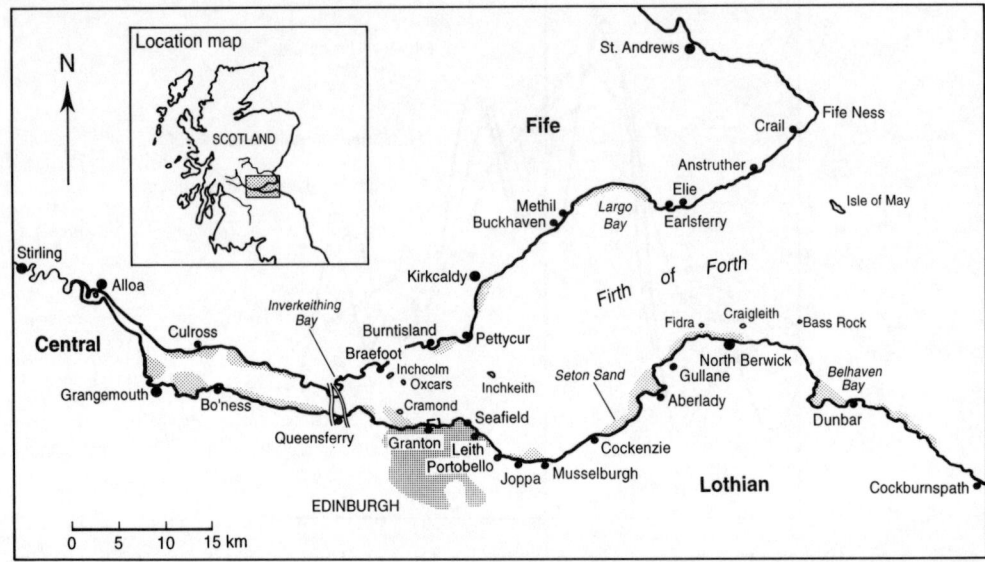

Figure 6.11 The Forth estuary and the Firth of Forth. Reproduced by permission of the Royal Society of Edinburgh and Dr D.S. McLusky.

The main features of this outer Firth of Forth are tidal currents and wind-driven currents. In this sense, as already mentioned, it is really part of the North sea oceanographically. The motion of the water tends to be inwards at the northern half of the mouth of the Firth and outwards at the southern half. The weak river outflow also hugs the southern shore. There are some distinctive features, namely, some stratification present for some months during the spring which can cause a reverse flow near the north coast (Fife). The confined nature of the domain compared with the North Sea also increases the importance of detailed monitoring of pollution.

From the modelling point of view, seasonal thermoclines are very difficult to simulate; the easiest approach is to assume a two-layer fluid (fresh water over saline water, or warm water over colder water). One of the consequences of allowing density to vary, is that this introduces *modes*. The zeroth-order mode is the homogeneous, constant density solution to the equations. The first-order mode is the solution that needs to be added to the constant density solution in order to produce the two-layer solution. Higher-order modes provide greater detail, in particular, more in the way of vertical structure. This, however, is the main modelling problem in introducing modes. Modes have very different characteristics in terms of typical time scales: strictly, the zeroth-order mode has a time scale that is very different from those of all higher modes. To illustrate this, the wave speed typical of a very long wave in homogeneous water is $\sqrt{(gh)}$, where g, the acceleration due to gravity, is about 10, and h, the depth of the estuary, is between 10 and 50 m (see Section 2.5). On the other hand, the wave speed in a two-layer sea associated with a disturbance on the interface, which might be an internal tide, is $\sqrt{(\rho_d gh/\rho)}$, where ρ_d is the *difference* between the densities of the two layers, and ρ is a typical density. The quantity ρ_d/ρ is typically 1/100, and hence wave speeds are very much slower. Wave speeds associated with higher modes are also much slower than that of the zeroth-order mode. If typical distances are equivalent, say, both are 10 km, a typical estuary width, then a direct consequence of this is that velocities associated with first and higher modes will be much slower than that associated with the zeroth-order mode. The modeller thus needs an efficient way of dealing with a motion that is evolving on two very different time scales. So-called *time-splitting* techniques are available that introduce two time scales such that numerical solutions can, schizophrenically, follow them both at once. This technique works only if the interaction between the two types of motion is zero. This becomes less true the more modes that are taken into account.

Hence detailed modelling of the seasonal currents that occur in the outer Firth of Forth is not an easy task. Recall Chapter 5, in particular the discussion of vertical structure in Section 5.3, where the zeroth-order mode dynamics are called *barotropic*, and first-order mode dynamics are called *baroclinic*. Because of the difference in time scales, it is possible for the Coriolis acceleration to have some bearing on baroclinic currents. The reason for this can be seen explicitly since the length scale (wave speed)/f (where f is the Coriolis parameter defined as 2 × the angular speed of the Earth × sin(latitude)) defines a length scale on which the rotation of the Earth exerts an influence on waves. This is called the Rossby radius of deformation. For wave speeds appropriate to baroclinic motion, this length is about 100 km, which would be appropriate for the North Sea scale and inappropriate for estuaries and inlets. However, for wave speeds one-tenth of this, as would be typical of baroclinic motion (two-layer or internal tide motion) this Rossby radius is called the *internal* deformation radius and is about 10 km. Now this obviously cannot be ignored in the Firth of Forth, which is 14 km wide. Indeed, it is this influence of Coriolis acceleration that is thought to be responsible for the, albeit small, freshwater outflow from the River Forth hugging the southern coast. To summarise, therefore, river flow is fresh water that rides over the bulk of the saline Firth of Forth water causing (locally) a two-layer regime. The baroclinic dynamics forces the stream to the right as it flows, which is the standard influence of the Coriolis acceleration. Hence the observed picture of a river flow along the Edinburgh side of the Firth of Forth.

In the western portion of the Firth of Forth, called the Forth Estuary, more typical estuarial dynamics apply. There have been one-dimensional models that have concentrated on modelling particular facets of the estuary's physics, chemistry and biology. The tides of the Forth Estuary exhibit a double high water, and a one-dimensional model incorporating a solution of the mass and momentum conservation equations, including variable depth, is able to predict this. If advection and diffusion (see Chapter 7) are included, such one-dimensional models can also predict the salient features of sediment transport or dissolved oxygen levels. However, such models are more than usually heavily dependent on starting observations and other boundary conditions. Indeed, it is possible to drive a 'model' entirely by the

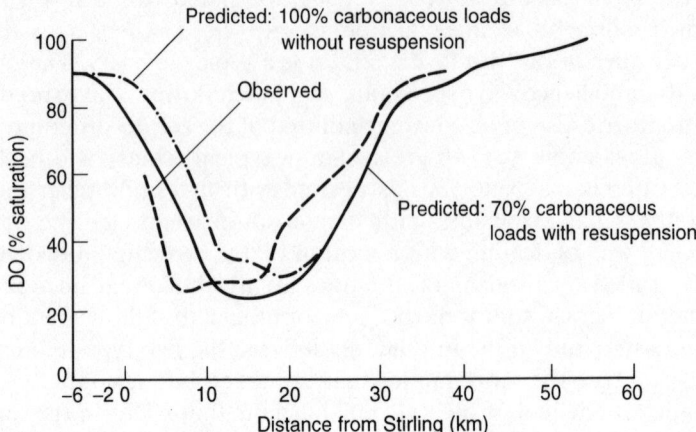

Figure 6.12 Observed and predicted dissolved oxygen (DO) distributions. Tidal range 5.2 m; river flow exceedance 80%. Reproduced by permission of the Royal Society of Edinburgh and Dr D. S. McLusky.

input of data. Such a 'model' is not really a model at all, but a database on which some simple dynamic rules have been superimposed. The first models that were built to describe the spread of effluent discharged into an estuary were of this type (see Chapter 7). Figure 6.12 gives some results from a one-dimensional model for dissolved oxygen levels compared with observations. The imperfections in the model can be ascribed to the model assumptions of a single source of organic pollution at the head of the estuary, and the constant cross-sectional area. Once the latter of these assumptions had been relaxed, the predictions improved, but no longer were there exact solutions available.

Chapter Seven

Diffusion modelling

7.1 Introduction

One of the clearest indications that there is human habitation on this planet is the presence of artificially produced material in the world's oceans, seas and estuaries. Sadly, this material is often poisonous to some degree. The name *pollution* has been coined to describe foreign often toxic material in the sea. Once pollution is present, it does not remain unchanged but is pulled and pushed around by currents present in the sea. Even if the sea is quiescent, a patch of pollution will almost always grow and spread, diluting as it does so. This process is called *diffusion*. Molecular diffusion can be observed if a grain of potassium permanganate (purple) is placed in still water. A purple patch gradually grows. Of course, this growth is enhanced if there are currents present. The school experiment that demonstrates convection by dropping a crystal of $KMnO_4$ (potassium permanganate) in a beaker of water being heated over a Bunsen burner clearly demonstrates this. Similarly, in the environment pollution can be made to disperse quickly in the presence of strong currents that are appropriately variable. A word must be said here about the use of the word *dispersion*. Quite rightly, dispersion is used as a synonym for diffusion. However, dispersion has a special meaning for physicists and applied mathematicians; it means the change (increase) in the wavelength of a wave (or group of waves) as it propagates. In view of possible confusion, therefore, the word dispersion will not be used as meaning general diffusion.

Although diffusion can and does take place at the same molecular level at sea as it does in the laboratory, far more important is the diffusion via turbulence. Turbulence, the random commotion of water, is ideal for diffusing pollution and it does it reasonably evenly in all directions. It is most important for the modeller to model the effects of turbulence when modelling diffusion, and since the effects of turbulence are the same as the effects of molecular diffusion only one thousand times greater, molecular diffusion can be safely neglected. Other effects, not in themselves diffusive, can greatly enhance or inhibit diffusion. At a shear, which is often present near a boundary, the unidirectional current varies from zero on the boundary itself to a large value only a short distance from the boundary. It is particularly efficient at diffusing any pollution. It is far more efficient, for example, than turbulence, which is suppressed near boundaries (solid ones at least). On the other hand, there are convergence zones in some flows (Langmuir circulation, for example; see Chapter 6). These convergence zones can act anti-diffusively, especially for buoyant contaminants. Their ability to re-concentrate hitherto dilute toxic substances is one of the main reasons for their study.

As far as modelling diffusion is concerned, one rather different species of model is the *particle tracking* model. Particle tracking or Lagrangian models are well suited to modelling the diffusion of particles because their *modus operandi* is to follow individual particles. Since the pollution can be simulated as a collection of marked fluid particles, these can be tracked by the model and various parameters (size of patch, location of its centre, etc.) output at appropriate stages. Let us look at some details.

7.2 Modelling the process of diffusion

To the fluid mechanist, diffusion is merely a consequence of applying the laws of fluid motion. Granted, they need to be applied with precision, so that effects mentioned above such as turbulence and current shear near boundaries are adequately represented, nevertheless diffusive effects should appear as consequences. To the practical modeller of real diffusion of real pollution this is not useful. In real life, too much is happening for everything to be included even in the most advanced of models. For example, an oil slick is one liquid (oil) interacting with another (sea water) in a complicated way. Even the oil is probably composed of several varieties which over time separate into liquids which vary from tar to light crude. Accurate hydrodynamic modelling of this is still very far from possible.

However, the fluid mechanist's view is useful in one respect. That is, diffusion is a turbulent process that happens once a parametrisation of turbulence is included in the dynamic balance. The simplest model of diffusion is one in which diffusive transfer of material (through action) is directly related to the gradient of the concentration. The constant of proportionality governs how quickly the diffusion occurs, and is called the diffusion coefficient. Such a model can be used to predict the diffusion of outfall material due to sewage or industrial waste spillage in a river, provided the spillage is large enough, homogeneous enough, and provided the river flow is reasonably uniform. This model is termed Fickean (after A. Ficke, who developed the idea in the mid-nineteenth century). Let us investigate the dimensions of such a diffusion coefficient.

If discussion is centred on the diffusion of momentum, the diffusion coefficient is our old friend eddy viscosity. The dimensions of the quantity, found in Chapter 2, are $L^2 T^{-1}$ (or perhaps density times this, $ML^{-1}T^{-1}$, which is the dynamic version). If, on the other hand, discussion focuses on a passive quantity, say temperature, salt or in fact any contaminant, then the amount of this passive quantity in a volume considered free of sources and sinks must be conserved. In its simplest form, this means that the time rate of change of the concentration of this passive quantity at any particular point must balance the spatial gradient of, not the concentration itself, but the agent that causes the diffusion of the concentration. In an ocean or sea, this agent is usually the turbulence-induced eddies. This agent is usually parametrised as being proportional to a constant times the concentration gradient (the Fickean assumption). The balance is thus

time rate of change of (concentration of passive quantity)

= gradient of [agent that causes the diffusion of the passive quantity].

In turn, the square bracket is a constant times the concentration gradient. If the passive quantity is labelled C, then in purely dimensional terms the balance becomes

$$\frac{1}{T}C = \frac{1}{L} k \frac{C}{L},$$

where k is the constant diffusion coefficient arising out of the Fickean assumption. Once again, therefore, it is seen that the dimensions of k are $L^2 T^{-1}$. This is now seen to be independent of the identity of the diffused quantity.

7.3 Some specific diffusion models

One of the most cited papers on diffusion was written by Akira Okubo in 1971. He, along with several other authors, recognised that oceanic diffusion occurs through many different mechanisms: turbulent eddies, shears of several origins, as well as biological agents. Obviously, it is not possible for a single model to simulate every diffusive effect. Okubo took on the task of examining many different measurements. In order to make some sense of all these data, it is useful (one could say essential) to provide a benchmark on which to measure and compare. Most experiments, certainly all those considered by Okubo, involve observing the spreading of a patch of dye. A patch of dye has a centre of mass, and a distribution of mass about this centre. All distributions have a variance which measures how spread out they are. As time progresses, the patch of dye increases in size, spreading out. It seems sensible to ask, therefore, what relationship there is between this measure of spreading and time. The straightforward Fickean diffusion, which is diffusion by turbulent eddies in the absence of shears, can be represented exactly by equations and solved. The result of this solution is that variance is proportional to time (the precise definition of variance is given in Chapter 9). Okubo's results give that variance is proportional, not to time t, but to time raised to the power 2.3, $t^{2.3}$. The graph from Okubo's paper is given in Figure 7.1

It can be seen from this graph that much of the data are scattered, and the line which corresponds to variance being proportional to $t^{2.3}$ is in reality a line of best fit or regression line. It will also be noticed that both axes are logarithmic which enable a power law to become a straight line. For those unfamiliar with statistical

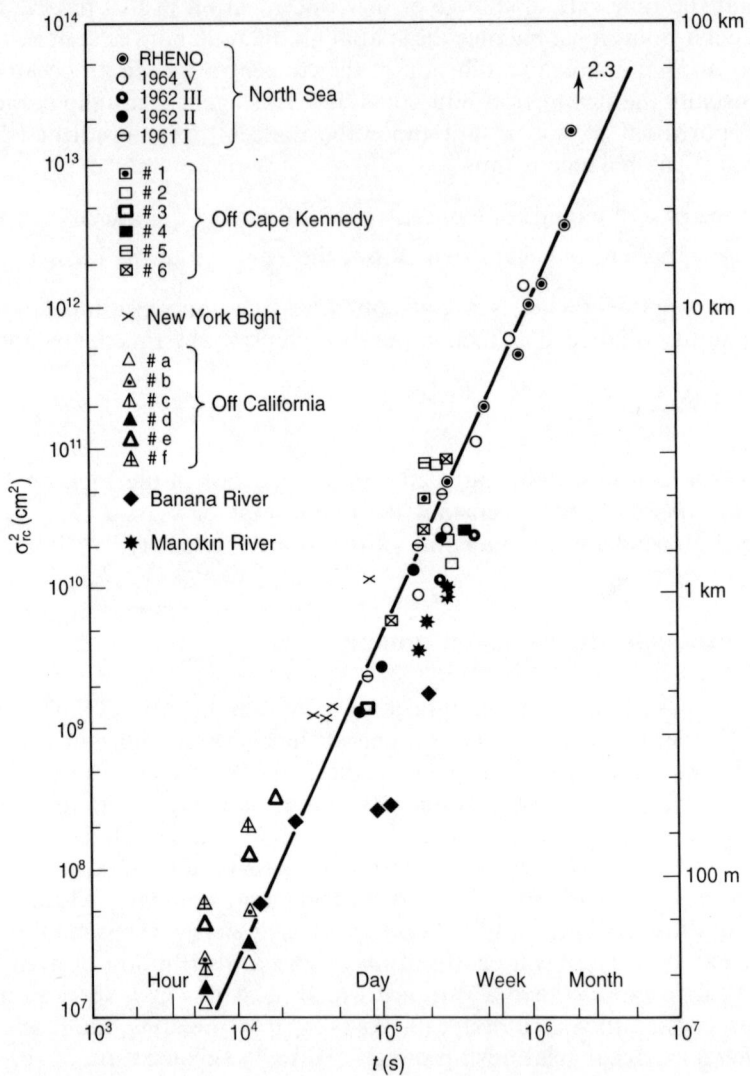

Figure 7.1 Diffusion diagram for variance versus diffusion time. From A. Okubo (1971). Reproduced with kind permission from Elsevier Sciences Ltd., The Boulevard, Langford Lane, Kidlington OX5 1GB, UK.

notions such as this, a summary can be found in Chapter 9. However, here we are more concerned with the mechanisms of diffusion and trying to model them. Figure 7.1 exhibits large scatter, even larger if one considers the logarithmic axes. However, the same axes show a vast range in length and time scales over which the power law is approximately valid. The validity of this power law for small scales is very open to question. Okubo shows no data for scales of 10 m or less, even those that are shown are very scattered around the 50 m length scale. It is therefore reasonable to deduce that the diffusive mechanisms are most distinct at these very different length scales, and Okubo's power law, although very seductive, is too much of a simplification of reality. A Fickean law will give variance proportional to t. There is another model due to Kolmogorov called similarity theory. Similarity theory arises from assuming that diffusion is homogeneous in all directions. This extra freedom allows material to diffuse more rapidly. In fact it can be shown mathematically that Kolmogorov's theory gives variance proportional to t^3. Okubo therefore reasons that the diffusion in the oceans is close to homogeneous, but constraints such as the ocean surface and perhaps the thermocline restrict the spreading. Some of the data he presents do locally, in fact, fit a t^3 power law. Okubo's $t^{2.3}$ power law is still accepted today as the best simple law of oceanic diffusion. It is time now to examine a different kind of diffusion model that makes use of modern computing power.

7.4 Diffusion models that track particles

In the 1970s, with the political crises in the Middle East and a general focus on saving energy, the world's attention on oil transportation by tankers intensified. Coincident with this, was an increase in public awareness of matters environmental. The final piece of the jigsaw was the unfortunate incidence of several oil spillages. Around the UK, the worst of these were the wreck of the *Torrey Canyon* in 1967 and, 11 years later, the break-up of the *Amoco Cadiz*. There were many lesser spillages between these dates. The early 1970s also witnessed the beginning of the computer revolution, making available a vast increase in computing power to modellers. The pressure to produce models of oil slick behaviour was therefore great.

A good method to model the behaviour of oil slicks, or any waterborne pollutant for that matter, is to hold the pollutant as a number of marked particles. The position of each of these particles is individually held in the computer, and the power of the computer is such that all these positions are held at each time step, being updated by the model. In this way the evolution of a patch of pollutant can be tracked. Commercially produced software began in the mid-1970s; for example, one called SLIKTRACK was produced by one of the major oil companies. Since then, more and more sophisticated oil spill tracking models have been produced. A relatively straightforward model, produced by Hunter (1980), is now outlined to give the general flavour of these particle tracking models.

Hunter's aim was to produce a computer model that could simulate the behaviour of an oil slick lying on the surface of the sea. Hunter also wished to develop computer software that could be used by an unskilled operator. To this end, the software would have an easy to use menu-driven front end so that anyone who had an interest in oil slick movement, but who was not perhaps familiar with the details of modelling, could input data and interpret output. This is now quite a common feature, but it was not in 1980.

When creating a model like this, one needs to decide what effects to include and what to exclude. This model predicted the movement of surface oil slicks, so the following factors were assumed to be important:

1. The tidal or non-tidal motion in the underlying sea water;
2. A wind-driven motion localised at the air–sea interface and caused by such mechanisms as surface water velocity, surface gravity waves and the direct action of the wind stress on the oil slick;
3. Spreading and mixing processes (e.g. gravitational and surface tension effects, surface wave activity and horizontal turbulence).

The underlying water movement, encapsulated in (1) and (2) above, causes the sea currents. These currents cause the patch of oil to move around as a whole, as well as to distort. However, the principal effect is to move the centre of mass of the oil slick. It is the spreading and mixing processes (3) that cause the slick to diffuse, i.e. to increase in size.

The tidal and other water movements are obtained, not by a model, but by interpreting observations. Other, later oil slick models have been entirely model based, using as their database the currents predicted from a primitive model of the type outlined in Chapter 6. This model however, simply uses a file of current data from Admiralty charts and other sources. In order to supply the wind-driven data (2) an empirical formula relating wind speed and direction is used. If a more sophisticated wind-driven current were to be introduced, some account would have to be taken of the rotational effects of Coriolis acceleration. Indeed, one of the suggested improvements by Hunter himself is to include some kind of Ekman effect. In this simple model, however, all such effects were omitted in favour of a direct drift that solely moves the centre of mass of an element in the direction of the wind but at some fraction of the speed. The name *element* here refers to one of the triangles into which the domain of the slick is divided to facilitate numerical calculation (see Chapter 3). It is the novel treatment of (3) that deserves most attention here. In order to simulate spreading by diffusion, Hunter incorporates a so-called Monte Carlo technique. This is a technique whereby a particular particle is moved randomly in a direction with a particular speed. Thus if one imagines a speeded-up film of a bird's-eye view of an oil slick composed of such particles, a kind of Brownian motion would be observed with individual particles dashing madly about, but with the whole patch growing steadily in size. This has a comforting feel of realism about it, but unfortunately only Fickean diffusion can be modelled by a random walk. It can be shown that under a random walk, the variance of the distribution of oil in an oil slick has to vary directly with time t and not, for example, with $t^{2.3}$, as Okubo's data indicate might be the case. Before returning to Hunter's model, it is worth pointing out that the seductive idea of being able to move particles around in a manner that seems so closely to mirror reality has been explored by several other researchers. Dyke and Robertson (1985) proposed a theoretical turbulence which was composed of randomly distributed and randomly sized eddies. They showed that the variance of a patch of oil in such a flow grows as time to the power of between 1.8 and 3. This is consistent with

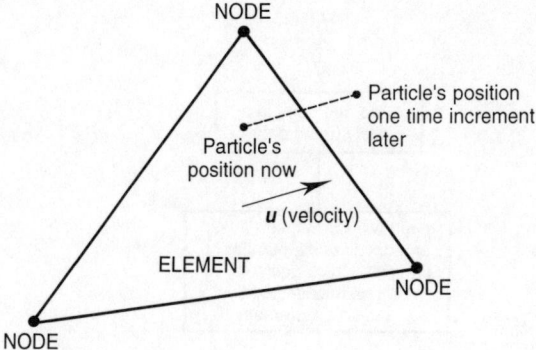

Figure 7.2 A typical element.

Okubo's result, but problems of initialising and calibrating such a model make it rather impractical for most purposes. Jenkins (personal communication) uses a model based on waves. The waves are random in direction with correlation times differing by factors of 2. Jenkins reproduced the Kolmogorov t^3 law for the increase in variance. It is true therefore that there remains much scope in models of this kind. Perhaps one day, such a model will give a true picture of turbulent diffusion. A cautionary note, however: complicated models that involve many parameters can, chameleon like, be made to simulate any set of data without improvement of understanding. The biological models of the next chapter demonstrate this all too well.

Let us return now to Hunter's model, and look at its operation. Hunter divided the domain into triangular elements (in the manner of conventional finite element methods), as shown in Figure 7.2. The particles of oil are tracked across the triangular mesh using an implicit finite difference scheme. The position of each particle is thus continually updated. The underlying velocity, due to tides (mostly), is taken as a constant throughout the element, as is the wind drift effect. Omitting the computational details, an analysis of the diffusive spreading process is given. The example of a radially symmetric patch after n applications of the finite difference scheme gives a variance of

$$\frac{(\text{speed} \times \text{time increment})^2 \times n}{2}.$$

Putting $t = n \times$ time increment and writing s as the distance travelled by a particle in one step gives variance $= ust/2$, as required by theory (and a Fickean model). Hunter's model also catered for the addition and removal of particles. It could not, however, cope with coastlines, or with oil that sank or differentiated into several different types of oil with different characteristics. Hunter's model was distinctive in that it was the first to incorporate the user explicitly, and that was its principal value. It is recognised that the model itself was too simple to be widely applicable. Here is a summary of model inputs and outputs:

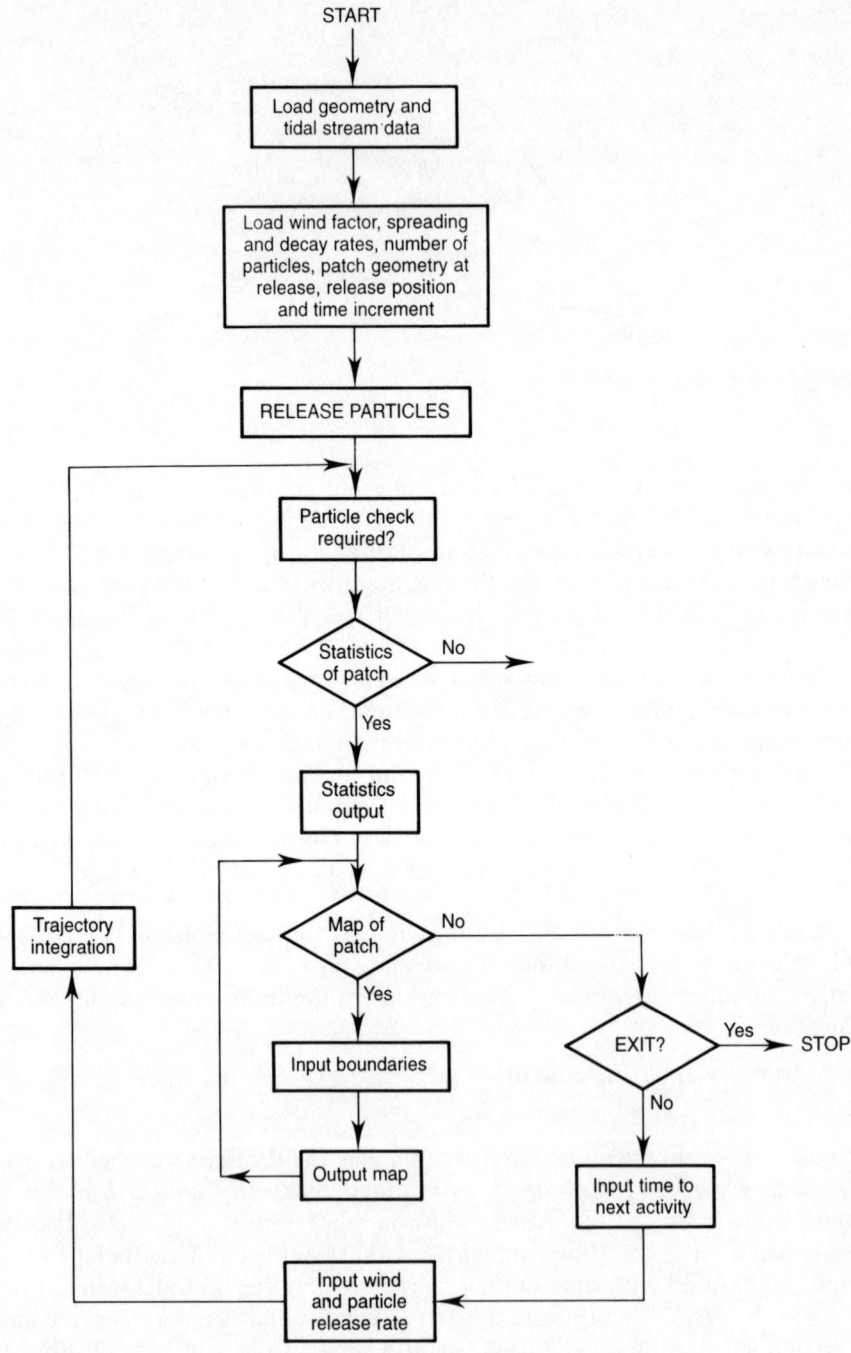

Figure 7.3 Flow chart for the oil slick prediction model.

Inputs
1. Wind drift factor (proportion and direction).
2. Diffusivity in SI units.
3. Decay constant.
4. Number of particles to be released initially (typical values would be 10 to 100).
5. Initial patch size, for instantaneous release.
6. Release position in latitude and longitude.
7. Time increment for computation (typically 900 s).
8. Time of initial release in GMT.
9. Whether particle checking is required within critical elements.
10. Elapsed time for next model output in hours, minutes and seconds.
11. Wind speed and direction for period defined in (10).
12. Release rate of particles from point source over period defined in (10).

Outputs
1. Number of particles.
2. Coordinates of the centre of gravity of the patch.
3. The direction of the principal axis of the patch.
4. The standard deviations parallel to and normal to the principal axis of the patch.

Initially, the patch is an ellipse of particles which is inputted under input (5). An additional map output with representative coastlines was also possible. A flow chart indicating the process is given in Figure 7.3.

7.5 Modelling diffusion in the North Sea

The motivation for the study of diffusion in the North Sea is the tracking and monitoring of pollutants. In the late-1980s, a collaborative European study (the North Sea Programme) brought together researchers from countries that border the North Sea with the object of furthering our understanding of the underlying processes governing the behaviour of the sea itself, the sediments and the life within the sea. One of the most important processes to get to grips with is diffusion.

The overall circulation pattern in the North Sea is reasonably well known and is free of controversy. There is an anticlockwise flow; southwards down the east coast of the UK, eastwards along the northern coasts of Belgium, The Netherlands and Germany, and then northwards along the west Danish coast, and finally along the west Norwegian coast to exit the North Sea (Figure 7.4). There is some 'leakage' across the North Sea in the form of the Dooley current, which is an intermittent feature flowing from the Scottish–Northumbrian coast, across the North Sea, towards the Skagerrak. The model itself is based on a finite difference discretisation of the type described in Chapter 6 (some details of the finite difference schemes themselves are given in Chapter 3). On top of the numerical model of water movements need to be added the inputs from the Firth of Forth, the Rivers Tyne, Humber and Thames in the UK, and from the Rivers Rhine/Meuse, Elbe, Schelde, Seine and Ems from

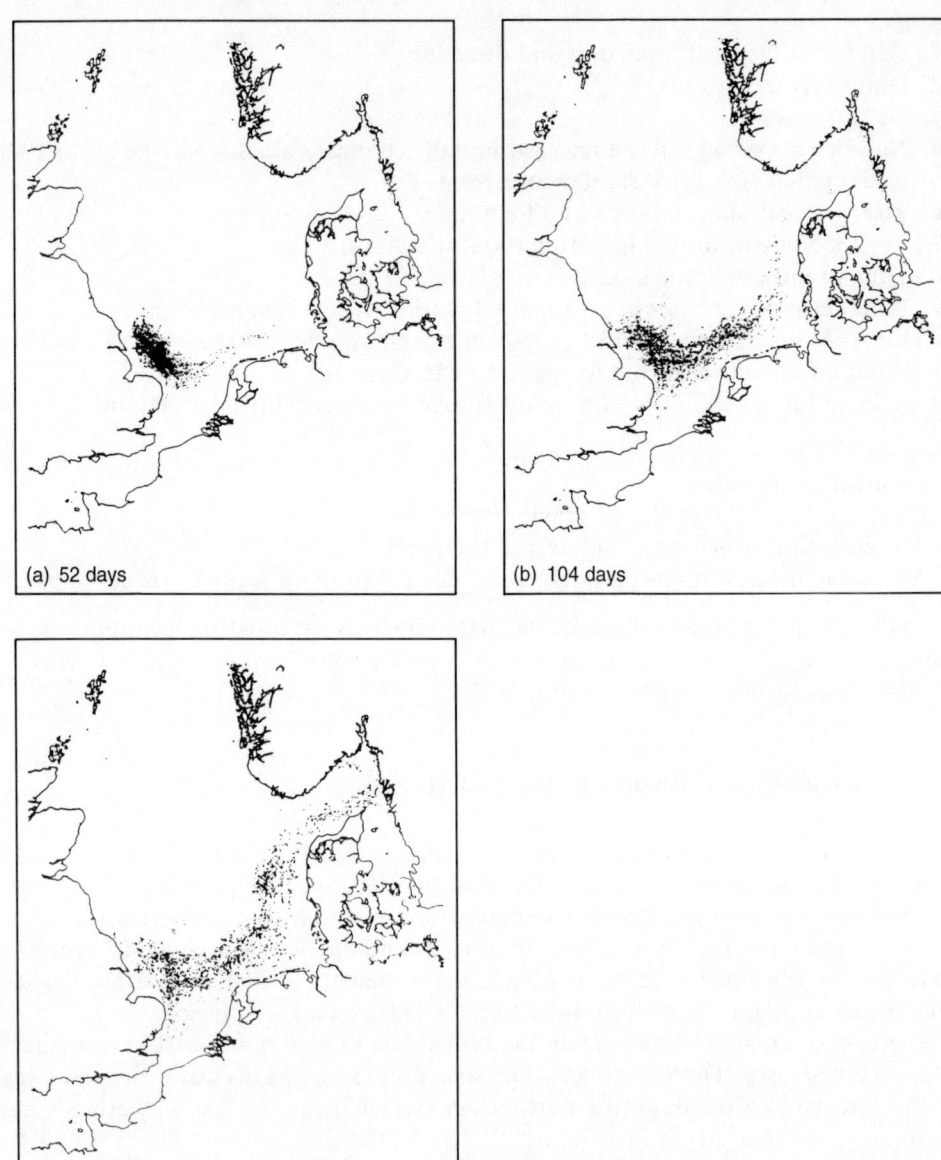

Figure 7.4 The diffusion of particles in the southern North Sea by currents and mixing.

continental Europe. These rivers often contain foreign material in the form of discharges (industrial waste) and runoff from farmlands containing fertiliser, which are carried into the North Sea as pollution. The North Sea, being a semi-enclosed basin, is particularly vulnerable to environmental stress (or environmental impact as it is known nowadays).

These discharges are assumed to be in the guise of particles. The model is based on solving the momentum equations but in a form that uses particle tracking (Lagrangian) techniques, in order that the pollution can be tracked explicitly. The horizontal resolution, by which we mean the spacing between the grid points, is about 20 km, and the time step is three hours. Two models are in fact used: a two-dimensional model with the stated resolution and time step, in which all quantities are integrated through the vertical (depth averaged), and a more sophisticated model which has 10 levels in the vertical and a much smaller time step of 12 minutes. The results from the two models are not significantly different. In both models, the time-dependent tidal current, not the mean, is used as the basis for computing the diffusion (spreading) of the particles, and in order to facilitate the handling of what would otherwise be too much data, the spreading simulations are vertically integrated. Spreading itself takes place within the model in two ways. The currents themselves exhibit turbulent fluctuations which by virtue of small eddies can spread passive contaminants that lie within the sea (as represented by the 100 particles). In addition, there are mixing coefficients in the model that purport to represent directly the turbulent diffusion process. Other diffusion processes that are not directly hydrodynamic in origin, such as those due to biological or chemical agents are not present in this model. There is an important distinction between the diffusion processes modelled here: *passive*, in which particles are carried around by the ambient flow (or perhaps at some fraction of the ambient flow); and biological and chemical processes which, although they may also be largely diffusive, are also *active* in that internal biology and chemistry can take place. Biological organisms can also propel themselves of course! This kind of active modelling is the subject of the next chapter.

Chapter Eight

Ecosystem modelling

8.1 Introduction

An ecosystem is an interconnected biological system involving animals, plants, nutrients and waste products. There are several features to ecosystems that make their modelling very distinctive. The essential feature of modelling as described so far in this text is the writing down of balances. These balances arise from physical laws such as the conservation of momentum (Newton's second law), the conservation of mass, and so on. For modelling ecosystems, there are no such laws. Instead, the understanding of how a particular component of an ecosystem behaves arises from understanding those effects that cause it to grow and those effects that cause it to die. Other effects, such as their direct interaction with another component, or the inclusion of a source of the component itself, can also be incorporated. Each component thus gives rise to an equation. On the left-hand side is the growth rate of the particular component, and on the right are the terms which influence its growth or its decay. In general therefore, we have an equation which has the following structure:

$$\text{growth rate of component } x = \text{(a positive constant)} \times x$$
$$+ \text{(a negative constant)} \times x$$
$$+ \text{source terms} + \text{interaction terms.}$$

The first term on the right-hand side represents growth factors (e.g. feeding, and growth due to internal metabolism of nutrients). The second term on the right-hand side represents the decay rate (e.g. defecation, dissipation by internal metabolism and ultimately death). The third term represents new sources of x (these might be conversions from other components or the mobility of x causing it to migrate into the domain of the problem). Finally, the fourth term represents the fact that what happens to other components can influence what happens to x. This single-variable model, although limited in its applicability, can model certain aspects of the growth of an organism quite successfully. Figure 8.1 shows what is called *logistic growth*.

Logistic growth is characterised by an initial exponential behaviour being limited by a ceiling or capacity. As such, although this limiting is deemed to be caused by

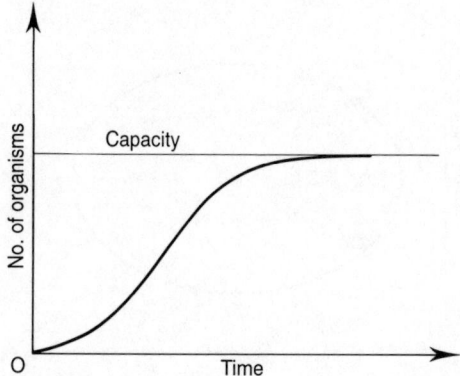

Figure 8.1 The logistic curve.

the organism itself – for example, the self-shadowing of an algal bloom in the upper layers of the sea – it could by caused for example by an (unmodelled) predator species. The fact is, the logistic model is a successful model of organism behaviour even though it is widely recognised that no organism exists in isolation, as such a single-variable model may at first imply. The next step from the logistic growth model is to include, explicitly, the predator species. For example, if x is the biomass or number of individuals of an animal, then obviously if its staple diet, y say, undergoes a drastic increase then x itself will increase. One thus may expect y to appear in an equation governing the growth rate of x, and indeed vice versa. We will meet this again when discussing the simplest ecosystem model, the predator–prey cycle. For now, we have established that each component has an equation, and we have also seen the general form of that equation.

What we have thus established is an equation for each component and within each equation a right-hand side that contains many parameters that govern the component's growth and decay rate, its sources (if any) and its interaction with its fellow components. It is quickly apparent that even the simplest ecosystem can have many equations. This is another distinctive feature of models of ecosystems: lots of equations, lots of so-called free parameters, but in most ecosystems models the equations themselves are linear and relatively easy to solve, certainly compared with equations that govern fluid flow. It is only recently that there has been recognition that even a single non-linear term, perhaps xy in the above notation, can lead to the kind of behaviour which has been christened chaos. This is largely unknown territory, even for the dedicated mathematician, let alone the marine scientist who is only a dabbler in modelling.

8.2 Predator–prey ecosystems

The simplest type of model is one that involves a single variable. This variable is usually 'population', and the name for this kind of model is a population model. A

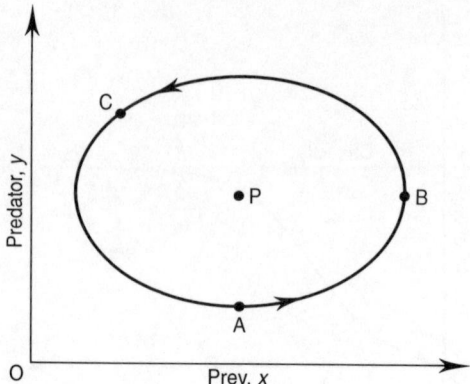

Figure 8.2 The predator–prey model, no damping.

population of a species can evolve in three ways, provided it is not entirely unpredictable. It can grow without limit, settle down to a steady value (perhaps a zero value) or it can exhibit periodic behaviour. The logistic behaviour mentioned in Section 8.1 is an example of a population that settles down to a value (its capacity). The right-hand side of the single equation that governs the change of population can be so structured that any one of these outcomes are possible. Mention ought to be made here of stochastic modelling, since population studies are its most straightforward manifestation. The word stochastic implies that there is a statistical or probabilistic element to the modelling. In population studies, several algorithms based on random walk techniques are possible. In these, there is a random element to the precise value of the population at any instant, and although averaging procedures can regain the previously mentioned underlying trends, only estimates of actual population are possible. In more sophisticated stochastic population models, the nature of the randomness itself can influence long-term growth or decay. More will be said about statistical modelling in the next chapter.

The simplest predator–prey model involves the interaction of just two species. The equation for each species takes a form that permits a closed graphical solution, as shown in Figure 8.2. It is instructive to look at this graphical solution alongside the equations themselves, which are written in the form

$$\text{Rate of change of } x = ax - cxy,$$

$$\text{Rate of change of } y = -by + dxy.$$

The prey (strictly, either the biomass of the prey or the number of individuals), labelled x, naturally grows (probably by eating grass, which is not included in the equations). The rate of this growth is represented by the constant a. On the other hand, the predator, labelled y, eats x at a rate governed by the magnitude of the constant c. The total rate of change of x is governed by these two competing effects. The predator naturally dies at a rate governed by the constant b (the death rate). On the other hand, it grows from eating x at a rate governed by the constant d. Note that although c and d represent the same process (y eating x) they are different because c denotes the effect on the prey, whereas d denotes the effect on the predator. For this simple model, arguments can be proposed that can justify putting $c = d$.

The cycling depicted in Figure 8.2 can be explained in words as follows: as the number of predators is low, to the left of point A on the curve, the prey can increase by grazing without fear of being pounced upon. As the number of prey reaches a maximum (point B), the population of predators also thrives due to the plentiful food supply. The inevitable consequence of this is a decrease in prey until (point C) they become scarce enough to diminish the predator population. Once the predator population reaches a low enough value, the prey thrives and the whole cycle begins again. This is the simplest model. If extra terms are added to the right-hand sides of these equations so that more sophisticated eating habits and more complex relationships between the number of predators, the number of prey and growth rates are represented, then it is possible for the curve to spiral inward towards point P. Such a point is called an equilibrium point and represents a stable point at which fixed numbers of predators and prey can live in perpetual harmony. When more variables are involved, a great variety of different stable, unstable and oscillatory states are possible.

Let us now examine one practical ecosystem model which helps to explain some of the well known characteristics of the biochemistry of the surface waters in the North Atlantic that were investigated during BOFS (the Biogeochemical Ocean Flux Study 1988–92, a UK national research programme funded by the Natural Environment Research Council and forming the UK end of JGOFS, the Joint Global Ocean Flux Study).

8.3 Modelling a real marine ecosystem

There are many biological and chemical signals in the ocean which one can attempt to model. The decision has been made here to look at a model which examines the role of phytoplankton in the carbon balance in the North Atlantic Ocean, particularly at the surface. This model is selected because of its relevance to global warming, one of the main concerns of today. Models which are published in the international scientific literature are very carefully researched, and usually stem from a history of similar models together with knowledge gained from many observations and experiments. What follows is therefore not a justification of all the modelling details in terms of their biology and chemistry. Instead, a documentary style account of the various balances is given, together with graphical results and a visual comparison with observations.

The model of Taylor *et al.* (1991) is essentially a point model. That is, only a single point in space is considered. There is no horizontal advection or transportation of material. The value of such models lies in their ability to isolate biochemical effects

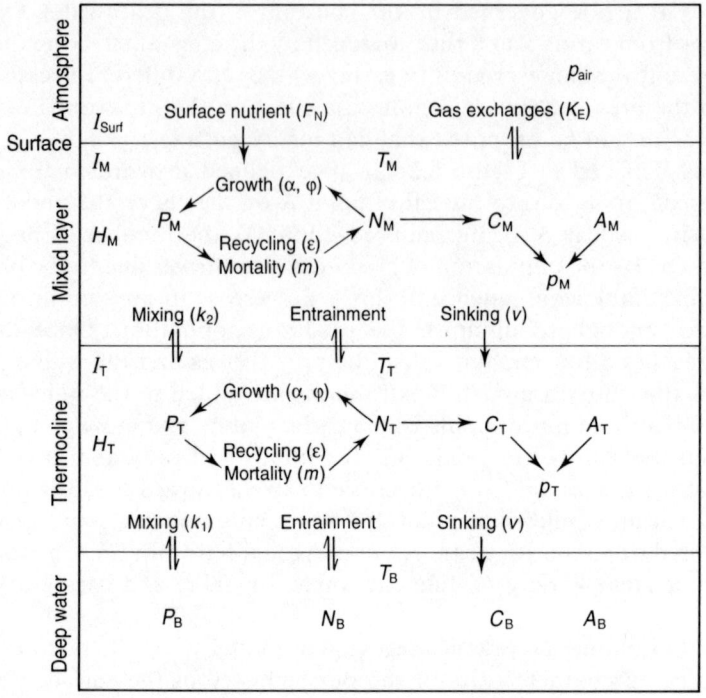

Figure 8.3 A marine ecosystem model. A, alkalinity; C, total inorganic carbon concentration; P, phytoplankton abundance; N, nutrient concentration. From Taylor *et al.* (1991). Reproduced with permission from American Geophysical Union, Washington, USA.

from hydrodynamic effects. Point models of biology and chemistry can be embedded in hydrodynamic models at a later stage, since only then can parameters appropriate to the interaction between fluid flow and biochemical effects be added. At the time of writing, very little work has been done on the interface between hydrodynamics and biochemistry in the sea. The model under the microscope here certainly contains no hydrodynamics. What it does consider, however, is some depth dependence. Although no horizontal structure is present, there is a mixed layer which interacts with the atmosphere above and a thermocline layer below. In turn, this thermocline layer also interacts with the deep water below it. The main biology and chemistry take place in the mixed layer and thermocline regions. In each layer, there are four variables (eight in all). These variables are phytoplankton concentration, nutrient concentration, concentration of total dissolved carbon and alkalinity. At its heart, the model therefore has eight rate equations which relate rates of change of these quantities to processes such as diffusion between layers, sinking rate of phytoplankton, carbon recycling efficiency, etc. None of the right-hand sides of these rate equations contains any non-linear terms, and hence there is no possibility that chaos will manifest itself. Models such as this are difficult to grasp from a cold start. Diagrams that

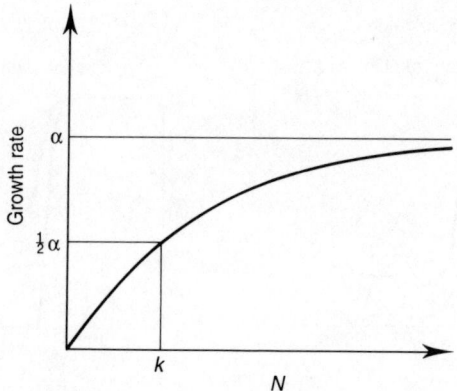

Figure 8.4 The Michaelis–Menten growth relationship.

illustrate the processes, such as that given in Figure 8.3, help us to understand what is taking place. For those who understand mathematics, this is a poor substitute for the precision of equations, but it is adequate for our purposes.

The most important input to a model such as this is nutrient, and it is thus crucial that this variable behaves realistically. Most researchers in this field are satisfied that nutrient obeys the Michaelis–Menten relationship. There is no rigorous justification for this in terms of well defined laws, although the following provides a justification of sorts. If N denotes nutrient concentration and α denotes a value of growth deemed to be an ideal maximum value, reached when conditions promote maximum nutrient uptake, then the growth of nutrient is governed by the expression

$$\frac{\alpha N}{k + N}.$$

This expression has the following essential properties. When there is no nutrient, there is zero growth. For very large N, the growth approaches the maximum (ideal) value α. For values in between, the growth is between zero and α, but it is always bigger for bigger N. Finally, the constant k is available for fitting to the data. In fact, if $N = k$, then the growth rate is $\frac{1}{2}\alpha$ (some readers may spot a loose analogy with radioactive half-life here). Figure 8.4 gives a general graph of the Michaelis–Menten relationship.

In the model of Taylor *et al.* (1991), nutrient limitation is represented by this relationship. The logic of this is seen once it is realised that phytoplankton grow by photosynthesis, and photosynthesis is directly proportional to irradiance. Therefore the presence of nutrient in the ecosystem will be limited by the amount of this sustaining irradiance. However, this global limiting via the Michaelis–Menten relationship will only apply to the total nutrient and at the microscopic level a more

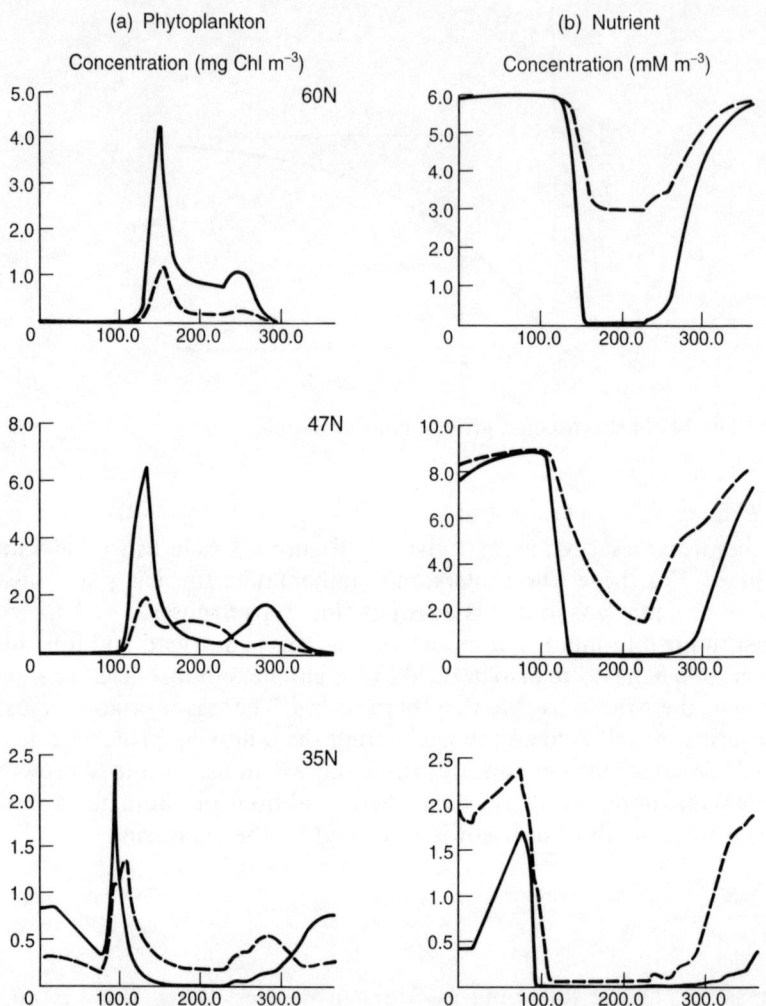

Figure 8.5 Model predictions of Taylor *et al.* (1991). In these runs the nutrient is diffusing up from depth. Solid lines, mixed layer values; dashed lines, values in the thermocline. From Taylor *et al.* (1991). Reproduced with permission from American Geophysical Union, Washington, USA.

complicated relationship governing the nutrient growth in each layer is appropriate. In this model, once the amount of nutrient reaches a certain level, the Michaelis–Menten limiting relationship dominates other effects.

The rate of change of phytoplankton in the mixed layer is governed by growth through nutrient absorption, a loss of phytoplankton through it sinking into the thermocline layer below, an additional death rate loss plus a diffusion, proportional to the difference between the concentrations of phytoplankton in the mixed and thermocline layers. Of course, this latter diffusion could act either as a source of phytoplankton (if the thermocline population exceeded the mixed layer population) or as a sink (if the reverse were true). In the thermocline layer, the negative of this diffusion process is present because the net effect of diffusion on the phytoplankton is zero, as we saw in Chapter 7; diffusion spreads out populations, but it cannot create or destroy them. In this thermocline layer, the phytoplankton concentration also grows through nutrient absorption and is depleted by mortality and through sinking. There is an additional diffusion via the deep layer which could be a source or a sink (as above) but is usually the latter since phytoplankton are not abundant in the deep sea. Below concentrations where Michaelis–Menten limitation is dominant, the nutrient concentration in each layer is governed by similar competing mechanisms. First, there is a recycling term due to the photosynthetic effect in the phytoplankton. This is subject to an efficiency index – after all the phytoplankton keep some nutrient to live, and only give up the surplus. Whatever nutrient diffuses into the mixed layer also diffuses out of the thermocline layer and vice versa. Diffusive interaction between the thermocline layer and the deep ocean also takes place. For the nutrient concentration, there are also inputs from the deep and from the atmosphere. These external inputs are crucial to this model since one of its prime functions is to predict seasonal changes; hence seasonal nutrient input is a prime driving mechanism. One of the main outputs of the model is total dissolved carbon. The model, besides tracking this dissolved carbon through its interaction with the growing, dying and sinking phytoplankton also includes a term which represents the transfer of carbon dioxide into (and perhaps out of) the sea surface. There are also the standard diffusion terms which take the same forms as previously described. The final variable is alkalinity, the value of which merely depends on the phytoplankton processes plus diffusion in each layer. In summary, therefore, the main dynamics of the model are phytoplankton growth, sinking and death, the interaction of this with nutrients, dissolved carbon and alkalinity, including all types of diffusive effects, and seasonal inputs of nutrient.

The key to whether a model such as this is a good one lies in how it compares with observation. Taylor *et al.* (1991) had the good fortune to be sitting on top of very good sets of data which had just emerged from cruises (1989) arising out of the Biogeochemical Ocean Flux Study. These data enabled the parameters of the model to be assigned with perhaps a little more than the normal confidence. Mention here ought to be made of the consequences arising from the absence of horizontal variation in the model. The model operates quite well at some latitudes, and fails miserably at others. The authors applied the model at several latitudes corresponding to where there was sufficient supply of data (models such as these are very data-hungry, unlike the hydrodynamic models of Chapter 6). They concluded that it was only appropriate to examine how the model performed at 60°N, at which latitude the phytoplankton reside primarily in the surface mixed layer throughout the year, and the main seasonal variations are in agreement with observations. Figure 8.5 shows the model predictions.

Fortunately, weather ship *India* was located at 59°N, 19°W for a number of years gathering data that can be used to help validate models such as this. The BOFS

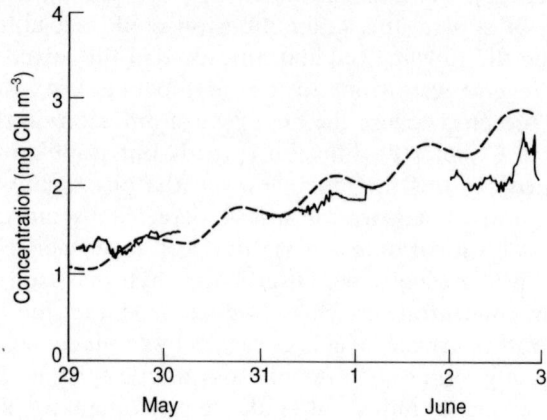

Figure 8.6 Chlorophyll-*a* concentrations. Solid lines, measured values; dashed lines, model predictions. From Taylor *et al.* (1991). Reproduced with permission from American Geophysical Union, Washington, USA.

cruises collected measurements of chlorophyll-*a* (a measure of phytoplankton concentration) and the partial pressure of carbon dioxide at 60°N, 20°W over a five-day period. The chlorophyll-*a* measurements could be compared directly with predictions (Figure 8.6). The measurements of partial pressure of CO_2 could also be compared with model output once the total dissolved carbon and alkalinity were related to this partial pressure via the standard chemical equations of sea water. The comparison is shown in Figure 8.7.

Another success of the model was the insight that was gained into coccolithophore blooms. Coccolithophores are a group of phytoplankton that have hard calciferous ($CaCO_3$) shells, are widespread in the oceans and provide a significant source of deep-sea carbon. Blooms of a species of coccolithophore called *Emiliana huxleyi* occur in mid-latitudes (45°–65°) in spring and summer, so much so that the ocean can appear milky white due to light backscattering. Taylor *et al.* used their model successfully to simulate various aspects of this coccolithophore bloom.

Let us now discuss the sensitivity and limitations of this model. The authors of this model, in recognition of the absence of lateral transfer in the model, added it crudely in one experiment via a simple northwards flow. Apart from an earlier spring bloom, there were no changes of major note. The sensitivity to changes in carbon dioxide exchange across the air–sea interface was examined. Some changes, particularly in summer values, were noted. Finally, the model was quite sensitive to the value of phytoplankton mortality and other parameter values which link phytoplankton abundance to nutrient.

It is quite typical of models such as this one to be quite sensitive to key parameter values. As the number of observations increases, so the estimates of the values of these key parameters can be improved. It must also be remembered, however, that many parameters (for example, the diffusion coefficients) are at best of doubtful

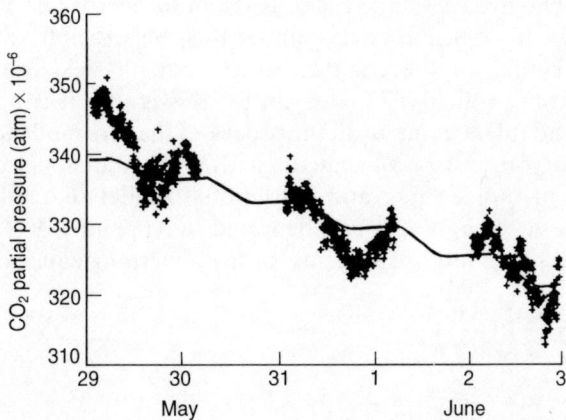

Figure 8.7 Partial pressure of CO_2. Crosses, measured values; full line, model predictions. From Taylor *et al.* (1991). Reproduced with permission from American Geophysical Union, Washington, USA.

pedigree and will always have a high degree of uncertainty associated with them. If a particular model prediction is very sensitive to such parameters, then the model output will be unreliable. When reading the details of an ecosystem model, always look for sensitivity analysis and arguments to back up the modelling predictions. These arguments should be based on a sound knowledge of the biological and chemical processes that are being modelled. A model which is given without its context is virtually worthless.

8.4 Other ecological models

The literature on ecosystems modelling is quite new in terms of modelling marine processes, but it is already extensive. There are many different kinds of approaches, approaches so different that it is often difficult for the lay reader to see any connection between them. This reflects the lack of an equivalent to the 'Newton's second law of motion' starting point to models of physical systems. One interesting new approach is to use simple equations, usually modified Lotka–Volterra, and plot solutions graphically. This approach, however, is limited to models that contain only at most three independent variables, so we will not dwell on details here except to say that there is a strong possibility of relating such models to modern developments in dynamic systems theory, which places no restriction on the number of independent variables (apart from the inability to clearly visualise more than three).

Instead, let us look at an attempt to model the deep chlorophyll maximum (DCM). Such a feature has been known about for at least 50 years, but it is only recently that its importance to the biological productivity of the region in which it occurs has been explicitly stated. The model detailed here is taken from work of Varela et al. (1994). Before detailing the biological model, we need to specify the physical model in which it is embedded. Unlike in the chapter on general physical models, some of the mathematics is given explicitly. The reason for this is that it is not possible to give a readily understandable account of all the processes that are modelled. If the equations are inaccessible to you, just recall that the hydrodynamic model takes into account the conservation of momentum, and a frictional model that takes the form of a turbulence closure scheme; details can be found in Appendix A. There is no need to include the troublesome non-linear terms, so the governing equations are of the form

$$\frac{\partial u}{\partial t} - fv = \frac{\partial}{\partial z}\left(\bar{v}\frac{\partial u}{\partial z}\right),$$

$$\frac{\partial v}{\partial t} + fu = \frac{\partial}{\partial z}\left(\bar{v}\frac{\partial v}{\partial z}\right),$$

where, as usual, u and v are the easterly and northerly components of the velocity, f is the Coriolis parameter, z is vertically upward, t is time and $\bar{v}$ is an eddy viscosity. It is this latter eddy viscosity that is treated rather differently from earlier models in that it is related to thermodynamic variables (instead of being assumed constant, as is usually the case). First, there are diffusion equations for salt and heat which introduce the standard diffusion coefficients (for simplicity these diffusion coefficients are assumed equal). The eddy viscosity is related to the turbulent kinetic energy (k) and the mixing length (l_m) via the simple relationship

$$\bar{v} = \tfrac{1}{2}\sqrt{k}\,l_m$$

which is due to Mellor and Yamada (1974). The turbulent kinetic energy is then related to the buoyancy produced by the presence of heat and salt through a number of quite straightforward equations which include the quantities the Richardson number, both the standard variety and the flux Richardson numbers, and the Brunt–Väisälä frequency. These terms are, hopefully, familiar from Chapter 2. The difference between the standard Richardson number and the flux Richardson number is that to obtain the latter, multiply the former by the thermal diffusivity divided by the eddy viscosity. The relationships

$$R_i = \frac{N^2}{(\partial u/\partial z)^2 + (\partial v/\partial z)^2},$$

where R_i is the (standard) Richardson number, and N is the Brunt–Väisäla frequency, given by

$$N^2 = -\frac{g}{\rho}\frac{\partial \rho}{\partial z},$$

relate these dimensionless numbers to the velocity and the density. The density is in turn related to the salinity and temperature through an equation of state. If some of

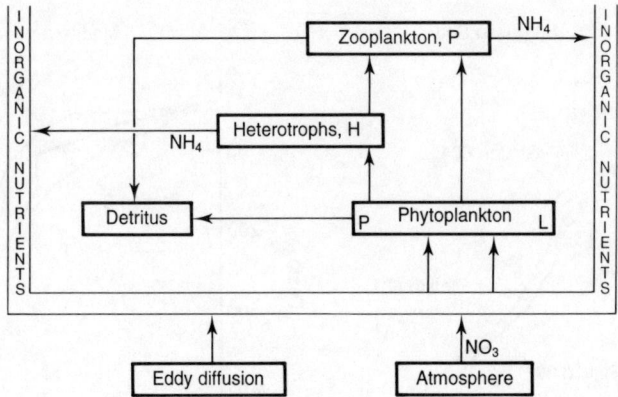

Figure 8.8 The trophic flow chart. From Varela *et al.* (1994). Reproduced by permission from the author.

the mathematical details above passed you by, the essential point to grasp is that the following biological modelling is soundly based on an up-to-date model of the underlying circulation. The biological variables used in this model are ammonia NH_4, nitrate NO_3, two kinds of phytoplankton (L for large and P for small), and heterotrophs, H (heterotrophs are creatures that get all their carbon from consuming green matter, in this case phytoplankton). The equations that they obey are of the usual variety, that is, they are rate equations with the rate of change of any particular variable being equal to (read 'governed by') uptake of the other variables, excretion and grazing. For completeness, and because we have avoided giving examples of the previous simpler model, we give these biological equations below. In what follows no understanding of any of the mathematical material will be assumed.

$$\frac{\partial (NO_3)}{\partial t} = \frac{\partial}{\partial z}\left(\lambda \frac{\partial (NO_3)}{\partial z}\right) - U(NO_3),$$

$$\frac{\partial (NH_4)}{\partial t} = \frac{\partial}{\partial z}\left(\lambda \frac{\partial (NH_4)}{\partial z}\right) - U(NH_4) + \psi_Z + \psi_H,$$

$$\frac{\partial L}{\partial t} = \frac{\partial}{\partial z}\left(\lambda \frac{\partial L}{\partial z}\right) - s_1 + U(L) - G_{ZL},$$

$$\frac{\partial P}{\partial t} = \frac{\partial}{\partial z}\left(\lambda \frac{\partial P}{\partial z}\right) - s_2 U(NO_3 P) + U(NH_4 P) - G_{HP},$$

and

$$\frac{\partial H}{\partial t} = \frac{\partial}{\partial z}\left(\lambda \frac{\partial H}{\partial z}\right) - s_3 + G_{HP} - G_{ZH} - \psi_H.$$

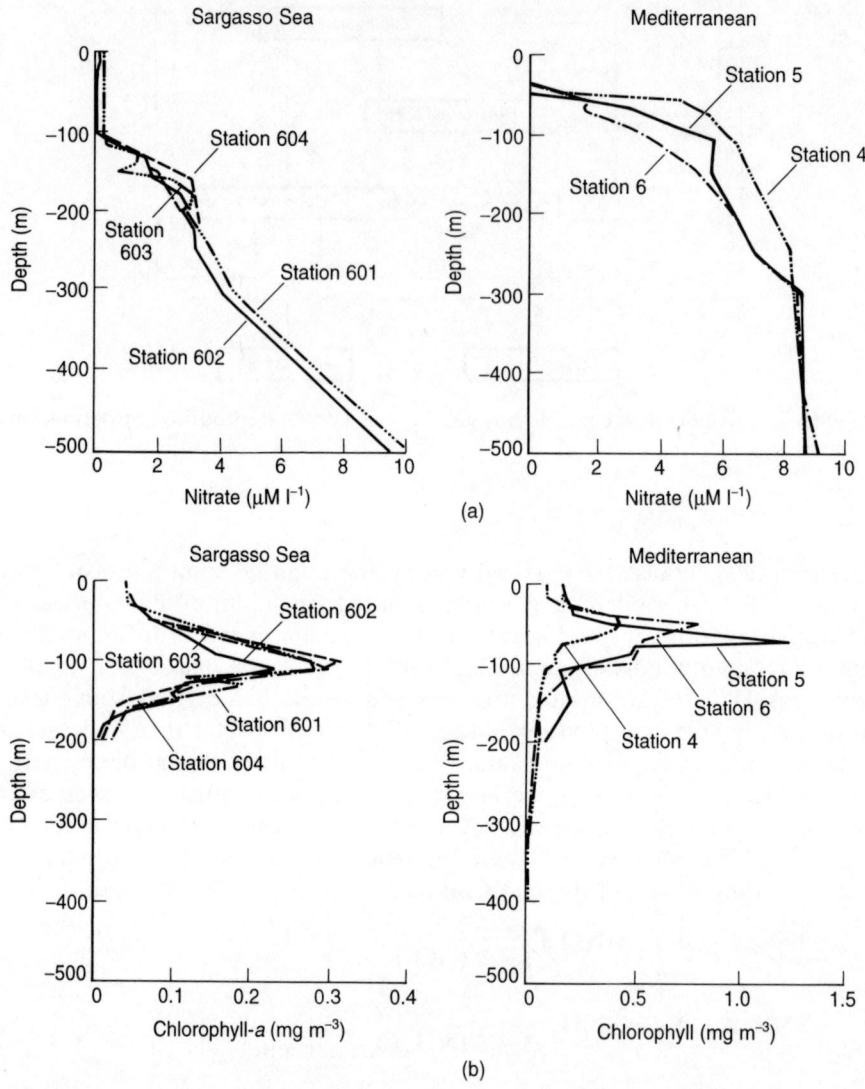

Figure 8.9 Model predictions: (a) nitrates; (b) chlorophyll-*a*. From Varela *et al.* (1994). Reproduced by permission from the author.

All these equations are, as we have said, rate equations. The left-hand sides of these equations give simply the rates of growth of nitrate, ammonia, large and small phytoplankton and heterotrophs, and the right-hand sides express how these rates are governed by various agents. The uptake is denoted by U, where the parentheses following indicate that which is uptaken. The composites NO_3P and NH_4P refer to

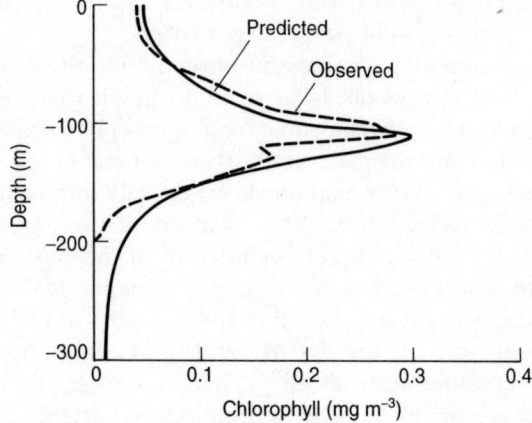

Figure 8.10 The chlorophyll variation with depth. From Varela *et al.* (1994). Reproduced by permission from the author.

uptakes of nitrate and ammonia, respectively by small phytoplankton. The terms s_1, s_2 and s_3 refer to the three sinking rates of large phytoplankton, small phytoplankton and heterotrophs, respectively. The G terms refer to grazing by zooplankton, heterotrophs and phytoplankton (Z, H and P, respectively), and ψ indicates ammonia excretion. The various uptakes are calculated through a knowledge of available light and available nitrogen using empirically derived relationships, which the authors freely admit are not universally accepted. The sinking terms are given by

$$s_1 = v_1 \frac{\partial L}{\partial z}, \qquad s_2 = v_2 \frac{\partial P}{\partial z}, \qquad s_3 = v_3 \frac{\partial H}{\partial z}.$$

Relationships such as the Michaelis–Menten formula that represents light limitation by self-shadowing are incorporated into the model through the uptake, sinking and grazing terms. The trophic model in terms of a flow diagram is shown in Figure 8.8.

Note that the presence of zooplankton is not modelled explicitly. The results of this particular model are quite good in that the predicted concentrations of nitrate seem to follow observations (see Figure 8.9). Of prime interest, however, is how the model predicts the deep chlorophyll maximum (DCM) which, after all, was the driving force behind the model. In order to see how chlorophyll enters the model, we note that light attenuation, I_z, at a depth z is governed by the expression

$$I_z = I_0 \exp(-(k_w + k_c + k_d)z),$$

where k_w is pure water extinction, k_d is the extinction due to factors such as organic matter and locally determined factors, but k_c is the all-important extinction due to chlorophyll-a concentration for which a linear function is assumed for simplicity. I_0 is the value for surface incident light. This model then is capable of predicting the distribution of chlorophyll-a with depth; Figure 8.10 shows one such profile using parameters appropriate to the southwest Sargasso Sea.

It can be seen that this model is successful in predicting the observed chlorophyll maximum. However, one always needs to remind oneself that models such as this are based to some extent on empiricism and have a number of so-called free parameters that can be tuned so that output fits certain observations. Good fit to data should therefore be only one factor that is examined. An equally important question to ask is whether the model correctly mimics other aspects of the physics, chemistry and biology of the sea. We have not given enough details of this particular model for the reader to answer this question, but in order for a paper to be published in the international literature, peer review should ensure that the model contributes a new insight into the prediction of the DCM, and that the modelling is accurate mathematically and credible scientifically. This model, combining as it does a turbulence closure scheme with chemical and biological variables is new and worthy of close examination by students of ecological modelling.

Chapter Nine

Modelling using a programmed text approach

9.1 Introduction

This last chapter provides the reader with the opportunity to do actual numerical examples of marine modelling using the models that have been introduced in the first eight chapters. Do not worry if you have no expertise in mathematics, or if some of the more technical parts of the book so far have seemed impossible to understand, the whole point of this chapter is to start from the very beginning and to take you through simple models using a step by step approach. In writing this, the author has been very influenced by the books of K.A. Stroud, who has over the last 25 years published programmed learning texts in the UK for engineering students. These texts, though not the favourite recommended books by lecturers and teachers of engineering students, have proved extremely popular with students, particularly those who struggle with the technical aspects of mathematics. Therefore what I wish to achieve here is a similar easy to follow run through some of the more elementary but nevertheless instructive marine modelling examples. As the problems are introduced, you are strongly advised to actually stop reading and do the problems *before* looking at the answer which will appear before the next part of the text. Although it is possible to give each problem a marine flavour, it turns out that in statistics in particular this often obscures the main point in that it makes what are quite simple principles seem complicated because of the nature of the details of the example. So although in what follows most examples are marine, this is not exclusively so.

9.2 Statistics

Statistics plays a very important role in marine science; one could even say a central role. The principal difficulty in writing a text such as this is to cater for the wide variety of previous statistical experience amongst the readership. The safest path to take is to assume very little previous knowledge. If what follows is too fast paced, then, certainly in the UK, there are a number of books that are designed for students in the last two years of compulsory schooling to cover those parts of statistics that appear in the National Curriculum. Generally, these have titles that contain the phrase 'GCSE mathematics' or 'Level 10 mathematics'. We shall start with the revision of what statisticians call *measures of central tendency*, which means ways of assessing where the middle of a set of data is. The simplest form of data is a list of numbers, although data are also often produced in the form of frequency tables. We shall deal with both.

Example 9.1

We wish to find the mode, median and mean of the following list of numbers:

5, 3, 6, 5, 4, 5, 2, 8, 6, 5, 4, 8, 3, 4, 5, 4, 8, 2, 5, 4.

First of all, do not worry about the definitions of these words; instead, we put the numbers in ascending order as follows:

2, 2, 3, 3, 4, 4, 4, 4, 4, 5, 5, 5, 5, 5, 5, 6, 6, 8, 8, 8.

The mode is the number that appears the most times:

mode = 5

The median is the number which is in the middle of the distribution:

median = 5

Finally, the mean of the numbers is the sum of the numbers divided by 20 (there are 20 numbers in all):

mean = 4.8

You should have obtained the answers: mode = 5, median = 5, mean = 4.8.

In this example, there is a clear mode since there are six 5's, and fewer of each of the other numbers (in general there is often a tie). There is an even number (20) of numbers, therefore the median is the average of the tenth and eleventh numbers. Since both of these are 5, so is the median. The mean is, uniquely, 4.8.

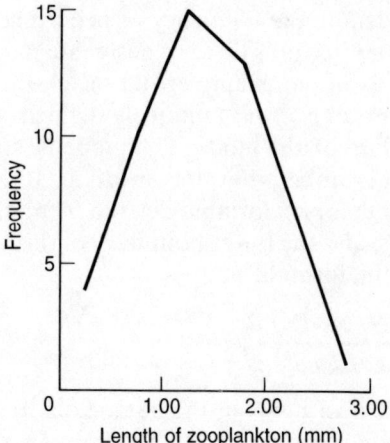

Figure 9.1 A frequency polygon.

Next, let us consider something a little more usual in scientific applications, that is, a situation where the numbers are grouped into classes and we have in effect a frequency distribution. This is usually given in tabular form. Table 9.1 gives the numbers of zooplankton of various lengths as measured by a marine biologist (adapted from research data and considerably simplified). The frequency polygon associated with these data is shown in Figure 9.1.

Table 9.1

Length of zooplankton (mm)	Number of zooplankton
0.01–0.50	4
0.51–1.00	10
1.01–1.50	15
1.51–2.00	13
2.01–2.50	7
2.51–3.00	1

The median of these data is still the middle number, but this is troublesome to find when the data take this form. In fact it is the value taken by the horizontal scale when a vertical line precisely divides the area under the frequency polygon into two equal halves. The mode is the peak of the frequency polygon. The mean is the quantity μ, which is given by the formula

$$\mu = \frac{\Sigma f_i n_i}{\Sigma n_i},$$

where the letters f and n denote the frequency of occurrence of the number, and the number, respectively. The subscript is there to designate that there are many numbers (i would run from one to six in our example) and the Σ sign denotes that summation over all i is to occur. The mean is always uniquely defined, although the same cannot be said for either the median or the mode. The mode is, straightforwardly, the class that contains the largest number, but the median either has to be determined graphically, or by a rather complex formula derived from its definition. If the median occurs in a particular class, and the lower boundary of this class is L, then the median itself is determined from the formula

$$\text{Median} = L + \left(\frac{N/2 - (\Sigma f)}{f_{\text{median}}}\right)c,$$

where N is the total number of items in the data, Σf is the sum of frequencies of all classes *lower* than the median class, f_{median} is the frequency of the median class, and c is the size of the median class interval. Given grouped data, it is easy enough to spot in which class the median lies; all the above formula represents is a mathematically precise way of dividing the area of this class to ensure that the median line so derived cuts the total area under the frequency polygon precisely in half. Now have a try at calculating the mode, median and mean:

mode = 1.255

median = 1.369

mean = 1.375

You should have obtained the answers: mode = 1.255 mm, mean = 1.375 mm and median = 1.369 mm. In this problem we meet several features that are typical in the handling of data. The mode is simply the mid-point of the interval (1.01–1.50) that contains the greatest number of animals. The mean follows by applying the formula remembering that in this instance the number of animals is multiplied by the length of zooplankton corresponding to the *middle* of the range (for example, 4×0.255 is the first entry in the numerator, 10×0.755 is the second, etc.). Finally, the median is calculated using the given formula with $L = 1.01, c = 0.49, \Sigma f = 14$ and $f_{\text{median}} = 15$. There are some more data sets that you can practice on at the end of this chapter.

Of course, in a brief summary such as this it is not possible to go into much detail in the way of statistical theory, nor is it desirable. The many specialist texts on statistics, having started as we have by introducing measures of central tendency, go on to discuss topics such as standard deviation, distributions, probability, and then to applied topics which include sampling, regression, hypothesis testing and experimental design. All of these have a role to play in marine science, but it would be over-ambitious to try to cover them in this book. Perhaps the most important point to make is that the central purpose of statistics is *inference*. The reason why data are analysed is to enable scientists to establish hypotheses (in a statistical sense) from the data. For this reason, the more theoretical aspects of probability theory are omitted here with the view that, should any be required it can be introduced *in situ*, as it were.

However, we do need to define variance and standard deviation. The variance of a set of numbers measures how spread out they are from their mean. It is defined by the formula:

$$\sigma^2 = \frac{\Sigma(X_i - \bar{X})^2}{N},$$

where the symbols have the following meanings: X_i denotes the data (i.e. the numbers themselves), $\bar{X}$ is the arithmetic mean, Σ is the summation sign which means that each number has the mean subtracted from it before it is squared, then the whole is divided by N, the number of numbers in the data set. The reason behind squaring each difference is that this makes all entries under the summation sign positive, hence making sure that the result of this sum is indeed a true representation of the spread of the data from the mean. Statisticians call this a 'measure of dispersion', but this is not an appropriate expression to use in a marine modelling book! In order to restore the dimensions, the variance is normally square rooted (hence the square on the left-hand side) and the symbol σ is called the standard deviation. Here is a practice example.

Example 9.2

Find the variance and standard deviation of the numbers

5, 3, 6, 5, 4, 5, 2, 8, 6, 5, 4, 8, 3, 4, 5, 4, 8, 2, 5, 4.

variance =   standard deviation =

You should have obtained the answers $\sigma^2 = 3.116$ and $\sigma = 1.765$. If you 'cheated' and used a calculator or a microcomputer, this is no problem as long as you are sure of what you have calculated and know what standard deviation and variance actually mean. The above answers only validate your arithmetic; they do not confirm your understanding! When a frequency table is involved, the definitions are of course the same but the method of calculation looks a little different. In fact, there is a very

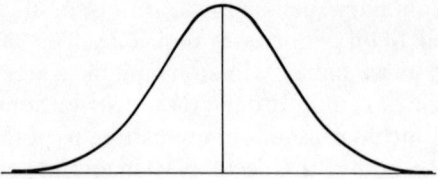

Figure 9.2 The normal distribution.

useful formula that can be derived from the definition of variance that proves useful in calculation. This states that the variance is given by the expression

$$\sigma^2 = \overline{X^2} - \bar{X}^2,$$

which can be read as variance equals the 'mean of the squares minus the square of the mean'. For grouped data, the following expression is the formula for variance:

$$\sigma^2 = \frac{\Sigma f n_i^2}{N} - \left(\frac{\Sigma f n_i}{N}\right)^2;$$

the standard deviation is of course the positive square root of the variance. Determine the variance and standard deviation of the data presented in Table 9.1:

variance = standard deviation =

You should have found that the variance is 0.408 mm^2 and that the standard deviation is 0.639 mm^2. Once again, the use of a calculator or software on a microcomputer eases the arithmetical chore, but this must not be at the expense of understanding. There are more examples upon which to practice at the end of the chapter. Let us now turn to applied statistics.

We have not dwelt on probability or distribution theory, since these topics are not central to our purpose, but once the mean and standard deviation are fixed, it is often acceptable to assume that data presented in the form of a frequency distribution approximate closely to normal. This assumption tends to be universal. In fact it tends to be assumed even when not appropriate. Users of statistical routines need to be aware of this. The first thing to remember is that the normal distribution assumes that the variation of frequency f with random variable x is the function

$$f(x) = \frac{1}{\sqrt{2\pi\sigma}} \exp(-(x-\mu)^2/2\sigma^2) \quad -\infty < x < \infty,$$

where σ is the standard deviation and μ is the mean. The general shape of this curve is shown in Figure 9.2. It is obviously desirable to have only a single distribution to cater for predictions based on the normal distribution. This has the consequence that all tables and software that contain information about the normal distribution have adjusted to a mean of zero and a standard deviation of one. Remember therefore to transform your data X into the normal variable z, sometimes called the z-statistic, through the simple transformation,

$$z = \frac{X - \mu}{\sigma},$$

before doing any statistical testing. The commonest of tests to use is the χ^2 test, which can be used to test whether or not a particular set of data fits a given hypothesis. A simple example is the toss of a coin. If a given coin is tossed 1000 times, say, and the outcomes are recorded, then this test can be used to decide whether the coin is biased or fair. Similarly, the χ^2 test can be used to decide whether or not data fit the conclusions drawn from a particular model. As hinted at above, however, one never gets *the* answer, and the criterion for acceptance or rejection of a hypothesis, to the applied marine scientist, is not God-given but is in fact dependent on assumptions involving the normal distribution.

Before we can do examples, we need to introduce the subject of hypothesis testing. This is the traditional first step on the road to *inference*, the main purpose behind most of statistics. Suppose we have some data, perhaps from observations taken on an oceanographic cruise. These data form what statisticians call a population. It is a collection of numbers arranged in a table or represented graphically. There will be certain statistics associated with the data – we have calculated the mean and standard deviation, but there are others. Now suppose further that we suspect that these data obey the form dictated by, say, the normal distribution. That is, we suspect that the mean and standard deviation conform to a certain normal, bell-shaped curve. We can use the χ^2 test to ascertain the truth of this hypothesis. This hypothesis is called the *null hypothesis* and is given the symbol H_0. If H_0 is rejected when in fact it is true, we say that a type I error has occurred. If we accept H_0 when it is actually false, we say that a type II error has occurred. Unfortunately, it is all too easy to make both sorts of errors, and it is always best to take a cynical look at the data, looking for oddities (*outliers* as statisticians call them) which may be due to human error in observing, or instrument failure, and which could distort the data and be the underlying cause of the type I or type II error. Finally, statisticians give the symbol H_1 to an alternative to the null hypothesis. Hopefully some of this will come alive through the next two examples.

The first of these examples is an introductory one involving that old standby, the tossing of coins; the second is a more practical example involving real marine data.

Example 9.3

Suppose a coin is tossed 1000 times, and the outcome is 530 heads and 470 tails. We might expect the outcome to be 500 heads and 500 tails, but then again it is the nature of chance that most of us would actually be surprised at such a precise obedience of the laws of probability. The pertinent question to ask is: is the coin fair? In other words, can the deviation from the ideal answer be attributed to chance, or is there a bias in the coin? In this case, the null hypothesis might be:

H_0: heads and tails occur with equal frequency.

We shall use the χ^2 test. In order to do this, we need an appropriate distribution. The χ^2 distribution can be found in Appendix B. In this table, the top row, labelled χ^2 which denote the *levels of significance*, gives a choice of thirteen numbers. These numbers represent significance levels so that, respectively, the columns that they head are appropriate to testing at the 99.5%, 99%, 97.5%, 95%, 90%, 75%, 50%, 25%, 10%, 5%, 2.5%, 1% and 0.5% levels. Let us choose the value 0.01, so that we are testing at the 99% significance level. Coin tossing is a process that has two possible outcomes (heads or tails); therefore the first row is chosen. The number in this row is

$$\chi^2 = 6.68.$$

You should have read the number 6.635. Now we calculate the value of χ^2 according to the formula

$$\chi^2 = \sum \frac{(\text{Observed} - \text{Expected})^2}{\text{Expected}}.$$

Remember, the summation sign is not a sum over the 1000 trials, but a sum over all possible outcomes. The calculated value is

$$\chi^2(\text{calculated}) =$$

The calculation should have proceeded as follows:

$$\chi^2 = \frac{(530 - 500)^2}{500} + \frac{(470 - 500)^2}{500},$$

so that $\chi^2 = 3.6$. This value is less than the value in the table, so we accept the null hypothesis H_0 and conclude that, at the 99% significance level, the coin is not biased. This is probably the correct conclusion, but if on examining the data we found 200 consecutive heads, we would want to research further into how the coin was tossed, etc. This latter point may seem a little silly here, but if we were dealing with real data, it is analogous to 'eyeballing' the figures and spotting if anything suspicious is present in the data. Mind you, one is much more likely to look if the hypothesis is rejected!

Here is a more realistic example. Table 9.2 gives the actual and expected values for catches of five species of fish.

Table 9.2

	Species A	Species B	Species C	Species D	Species E
Expected catch	25	5	7	31	35
Actual catch	20	4	17	26	30

First, state the null hypothesis for this problem:

H_0 states that:

You should have written 'H_0 states that the expected catch and the actual catch are the same'. Using a χ^2 test with parameter 0.01, do we reject H_0? To answer this we calculate χ^2 from the formula and get the appropriate value of χ^2 from the table in Appendix B:

χ^2(calculated) = χ^2(table) =

From your calculation and from the table, you should have obtained the values

χ^2(calculated) = 17, χ^2(table) = 13.3.

On the face of it, these results indicate that we should reject the null hypothesis. However, if we glance at the table of data, there is a very large discrepancy between expected and actual catch for species C. Without species C data, H_0 would have easily been accepted. The correct conclusion to draw therefore is that the figures for species C need to be re-examined and the reason for the glut of fish or the serious under-estimation of the catch ascertained. In passing, note that for an n-variable problem ($n = 2$ for the coin, and $n = 5$ for the fish) we look at the line $n - 1$ rather than line n in the χ^2 table in Appendix B. The reasons for this are rather technical, but the case $n = 1$ is worth some thought and gives a clue as to why.

The final topic to cover in this briefest of excursions into statistics is fitting lines to data. The most common example of this is the regression line, which is a line of best fit through a set of data points. Given a scatter plot as shown in Figure 9.3, there is a quite straightforward procedure for drawing a line of best fit through the data. An arbitrary line is drawn, then the square of the perpendicular distance of each point from this line is calculated. These are all added together, and the minimum

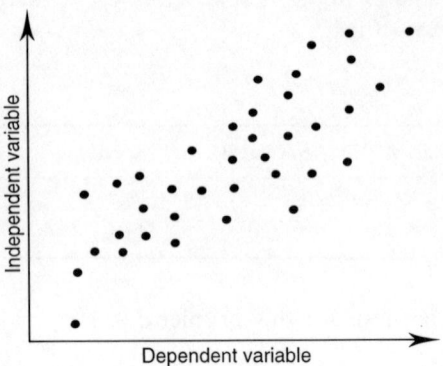

Figure 9.3 A scatter diagram.

value of this is found. The parameters of the line that correspond to this minimum value give the line of best fit. Difficulties arise only when the data are so scattered that there is virtually zero correlation, in which case the line of best fit has no meaning. In fact, there are always *two* regression lines. If x and y denote the standard axes, these regression lines are called 'y on x' and 'x on y', and if there is no correlation then these two regression lines are at right angles to each other. Recall that the term correlation refers to the measure of agreement between two sets of data. A correlation of 1 denotes perfect agreement, a correlation of -1 denotes perfect disagreement (for example, the rainfall at one point of an estuary, and the salinity of the water at the same point; as the rainfall increases, the salinity decreases and vice versa) and a correlation of zero denotes no relationship at all. The other complication occurs when there is obviously a relationship between two variables, but this relationship is not a linear one. This takes us into the heady topic of log–linear and log–log plots. These days most schools seem to have abandoned logarithms because they are no longer of any practical use as a calculating tool. They have gone the way of the *ready reckoner* and the *slide rule*. Unfortunately, logarithms have another function that has not been superseded – they are used to represent data where some kind of exponential growth is taking place. Those students that need to know about such things, such as students of biology and marine science, are thus faced with logarithms for the first time. There is, fortunately, no need to dwell at length on the properties of logarithms; all that is necessary is given below.

If an animal is growing exponentially, then its weight w might be related to time t through a relationship such as

$$w = a + b \exp(ct)$$

where a, b and c are known constants that are fixed once the species and its environment are also fixed. Of course, this growth will stop, but this is a simple illustration only. If we wanted to make t the subject of this formula, then we would subtract a from both sides before taking logs to obtain

$$t = \frac{1}{c} \ln\left(\frac{w-a}{b}\right),$$

where the symbol ln denotes the natural or Naperian logarithm. This particular logarithm function is the inverse of the exponential function, and is the 'log' referred to in the phrase 'log–linear' as in graph paper. We have still not given the reasons for needing to know about such graph paper. Well, take the expression just derived,

$$t = \frac{1}{c} \ln\left(\frac{w-a}{b}\right).$$

If data corresponding to $(w-a)/b$ were to be plotted on one axis of log–linear paper, and data corresponding to t be plotted on the other, then provided w and t were related in the way dictated by the above equation, the plot on the log–linear paper would be a straight line (with slope c). Once a scatter plot can be assumed to contain within it an implied linear relationship, then all the regression methods developed for straight lines can be brought to bear on the data. Table 9.3 gives some examples of relationships and the paper that should be used to display them as a straight line. In what follows, X and Y are the independent and dependent variables, respectively.

Table 9.3

Equation	Straight line	Description
$Y = \dfrac{1}{a + bX}$	$\dfrac{1}{Y} = a + bX$	A hyperbola: use ordinary graph paper
$Y = ab^X$	$\ln Y = \ln a + X \ln b$	An exponential curve: use log–linear graph paper
$Y = aX^b$	$\ln Y = \ln a + b \ln X$	Geometric curve: use log–log graph paper
$Y = \dfrac{1}{ab^X + c}$	$\dfrac{1}{Y} = ab^X + c$	Logistic curve: use log–linear graph paper (with care!)

It is possible to obtain commercially special paper that renders variables that are related logistically (the last entry in Table 9.3) as a straight line. However, log–linear paper can be used provided the equation is transformed into exponential type by treating $(1/Y) - c$ as a variable.

One important question we have not yet addressed is how to assess whether or not a particular law is suitable for a given set of data; we cannot always rely on

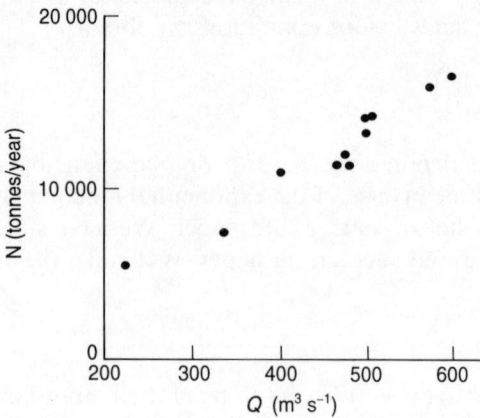

Figure 9.4 Total nitrogen against river flow.

simply 'eyeballing' it. If we wish to compare two sets of figures in a quantitative manner, then we calculate a correlation coefficient. There are several such coefficients to choose from, but the one most commonly used is the Pearson correlation coefficient, which is 1 for perfect agreement, -1 for perfect disagreement, and 0 for no relationship at all. To calculate the Pearson correlation coefficient, r_{XY}, we should use the formula

$$r_{XY} = \frac{(1/N)\Sigma[(X_i - \bar{X})(Y_i - \bar{Y})]}{S_X S_Y},$$

where

$$S_X^2 = \frac{1}{N-1}\Sigma(X_i - \bar{X})^2 \quad \text{and} \quad S_Y^2 = \frac{1}{N-1}\Sigma(Y_i - \bar{Y})^2,$$

and all summations are over all data points. The presence of $N-1$ rather that N in some of these expressions may perplex some readers, but it is a consequence of various statistical assumptions, in particular the assumption that the variates when standardised to a mean of zero and a standard deviation of one obey a slightly distorted normal curve called a *t-distribution*. Although the above formula gives the definition of r_{XY}, we give below the most widely used practical formulae for calculating not only r_{XY} but also the regression line of Y on X in the form $Y = AX + B$. Purists will notice a missing factor of $(N-1)^2/N^2$ in the formula for r_{XY}, but this quantity is very close to one in most practical examples. In fact, if it is not, then any straight line drawn through such sparse data has only scant value.

$$r_{XY}^2 = \frac{(N\Sigma XY - \Sigma X \Sigma Y)^2}{|N\Sigma X^2 - (\Sigma X)^2||N\Sigma Y^2 - (\Sigma Y)^2|},$$

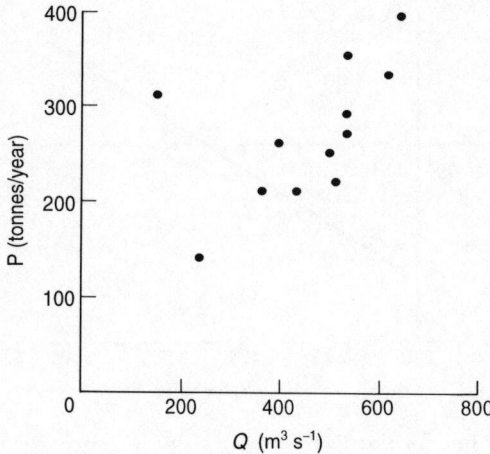

Figure 9.5 Total phosphorus against river flow.

$$B = \frac{N \Sigma XY - \Sigma X \Sigma Y}{N \Sigma X^2 - (\Sigma X)^2}, \qquad A = \frac{\Sigma Y - B \Sigma X}{N}.$$

Let us now do an example.

Example 9.4

Table 9.4 gives the discharges of total nitrogen (N) and total phosphorus (P) through the River Göta in tonnes per year, as measured in the years 1972–82 (inclusive). The quantity Q denotes the river discharge in $m^3 \, s^{-1}$.

Table 9.4

Discharge	1972	1973	1974	1975	1976	1977	1978	1979	1980	1981	1982
N (t/yr)	13 600	8900	13 500	12 700	7000	16 700	14 900	13 600	18 700	18 000	16 400
P (t/yr)	310	210	250	220	120	350	290	310	390	330	270
$Q \, (m^3 \, s^{-1})$	150	365	505	515	240	535	535	435	645	620	535

First, we need to plot the two scatter diagrams of the discharges of nitrogen and phosphorus. These are shown in Figures 9.4 and 9.5, respectively. The variable Q is the independent variable, and it is seen that the data are suitable for a linear regression line to be appropriate. Calculate the two correlation coefficients r_N and r_P:

$r_N = \qquad\qquad r_P =$

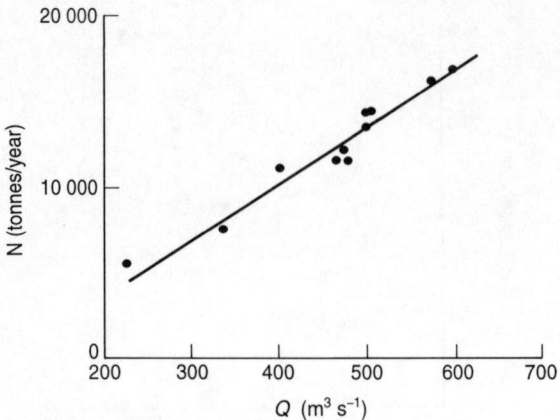

Figure 9.6   Regression line for nitrogen.

Whether you used the formula directly, used a calculator or software, you should have obtained the answers $r_N = 0.73$, $r_P = 0.53$. Although both correlations are positive, they are not particularly high, so it is not obvious that linear regression is the best way to obtain reliable predictions. One may find a better non-linear relationship, but looking at the scatter plots does not immediately suggest any obvious alternative candidates. We therefore still press ahead and calculate the linear regression lines, but bearing in mind that predictions need to be treated with some caution. It is possible in fact to place error bars on the value(s) of r_{XY}, but such refinements are considered outside the scope of this introductory text. Try calculating the two regression lines for these data using the formulae:

For nitrogen: $N = A_N Q + B_N$ $A_N =$
$B_N =$

For phosphorus: $P = A_P Q + B_P$ $A_P =$
$B_P =$

You should have obtained the answers

$A_N = 17.06$, $B_N = 6121.06$,
$A_P = 0.258$, $B_P = 158.22$.

Note that no attention has been paid to units here. The data are given in a mixture of units, as is quite typical (one might even say prevalent) in marine science with its

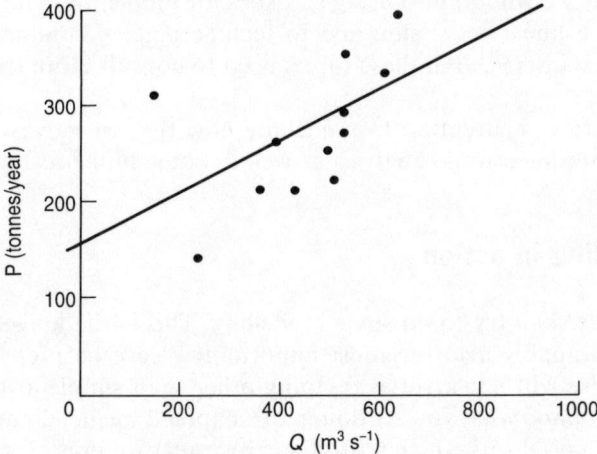

Figure 9.7 Regression line for phosphorus.

long nautical traditions, and no conversions to, say, standard SI units have been made. This may annoy the purists, but in the calculation of lines of best fit, the geometric distance between the data points and the regression line has been minimised and this process is independent of units. Only if we wish for sensible units for the constants A_N, A_P, B_N and B_P does it become necessary to standardise.

Finally, in this example, let us do some predicting. Use the regression lines to predict the values of nitrogen and phosphorus in the River Göta when the river discharge is 800 m³ s⁻¹.

N =   P =

Either from drawing these lines on the graphs (shown in Figures 9.6 and 9.7) or, more accurately, from inserting $Q = 800$ into the formula for each line in turn, you should have obtained the answers

$$N = 19\,769 \qquad P = 364.$$

The first, less accurate method is acceptable, particularly because it keeps you in touch with the data, reminding you how scattered the points are, and hence how low the correlation is. Most importantly, it indicates how much (little?) reliance can be put on these predictions. A correlation of 0.9 would be considered a reasonable figure, and the data fall well below this.

This is as far as we will go into using statistics for modelling. The next step would be to look at non-linear regression and to include placing confidence intervals on predictions. Those interested in these topics need to consult more specialist statistics texts.

Let us now turn our attention to modelling how the sea moves. We shall utilise our knowledge of dimensional analysis as well as some numerical methods.

9.3 Modelling in action

In this section, we shall try to do some modelling. This is the longest section in this book, but it is arguably also the most important. Please be prepared to do some calculations. These will not involve anything other than simple arithmetic, but you will need to understand what you are doing. In Chapter 2 we met dimensional analysis. This is a device that tells us which terms (i.e. processes) are important and which can be neglected in a given oceanographic situation. The first example involves the Gulf Stream.

Example 9.5

An ocean current of 1 m s^{-1} is 200 km wide. It moves in a path with a radius of curvature 1000 km. Calculate a Rossby number if the latitude is $40°$N. Discuss the modelling implications but ignore friction entirely.

The first step is to write down dimensions that are appropriate to speed U, length L, and Coriolis parameter f, since these make up the Rossby number $Ro = U/fL$. From the question statement, we can estimate these parameters:

$$U = \qquad L = \qquad f =$$

Deciding upon a typical speed is not difficult since $U = 1$ m s^{-1} is given in the question, and so is the only sensible candidate. On the other hand, there are two choices for a typical length. $L_1 = 2 \times 10^5$ m is the cross-stream dimension, and $L_2 = 10^6$ m (= radius times one radian of angular measure) is a measure of length associated with the curvature of the Gulf Stream itself. The Coriolis parameter $f = 2\Omega \sin(\text{latitude})$, where $\Omega = 7.29 \times 10^{-5}$ rad s^{-1} is the angular speed of the Earth is not controversial. It gives $f = 10^{-4}$ s^{-1}. From these figures, the two possible values for the Rossby number are

$$Ro_1 = \qquad Ro_2 =$$

Using the two possible values of L, the two values of Ro are 0.05 and 0.01, both of which are small enough for the advective terms in the equations of motion to be ignored. In fact, the Gulf Stream can be analysed using a linear theory unless bends in the stream or the narrowness of the stream itself lead us to choose $L = 10^4$ m, in which case the model has to be non-linear. This is entirely consistent with present theories in which non-linear terms are only important when relatively small-scale features are coupled with high current speeds. The power of dimensional analysis is clearly demonstrated, but also its principal weakness is that we now have to actually *solve* the equations, which dimensional analysis cannot help us to do.

Example 9.6

In this example, let us consider a coastal sea where friction is important. There is a tidal current of magnitude 1 m s^{-1}. The sea level rises 1 m and appreciable differences in the horizontal occur over distances of 100 km. The latitude is 56°N, and we assume a quadratic friction law of the form

$$\text{friction} = \rho C_D \boldsymbol{u} |\boldsymbol{u}|,$$

where ρ is the density, C_D is a drag coefficient of magnitude $\sim 10^{-4}$ m^2, and $\boldsymbol{u}$ is the fluid velocity just above the sea bed. Determine the magnitudes of the terms in the equation of motion,

$$\frac{\partial u}{\partial t} + u \frac{\partial u}{\partial x} - fv = -g \frac{\partial \zeta}{\partial x} + \frac{\text{friction}}{\rho},$$

and determine what the magnitude of the horizontal gradients have to be in order for the term $u(\partial u/\partial x)$ to become important.

For this problem, typical dimensions are

$U =$	$L =$	$T =$

We are given directly that $U = 1$ m s^{-1}, but what is L? In the vertical, a typical length might be 1 m, but this model does not include vertical dynamics (no variation with z, the vertical coordinate). So we deduce that L must represent a typical horizontal length scale which, from the statement of the question is 10 km or 10^5 m (compatible units are important here). Assigning T has the potential for being awkward since there are two naturally occurring time scales: one is the Coriolis parameter, and the other is the frequency of the tide. In mid-latitudes, the Coriolis parameter is $\sim 10^{-4}$ s^{-1}, and the frequency of the tide, corresponding to the M_2 semi-diurnal tide is

$$f_{M_2} = \frac{\pi}{\text{day}} = 1.4 \times 10^{-4} \text{ s}^{-1}.$$

So, fortuitously, T is of magnitude 10^{-4} s^{-1} satisfies both time scales. It is relevant to ask what would have been done if the time scales had been different. Which one would we have chosen? The answer is the tide, since it is the dynamics of tides that interests us in this problem (in fact, the Coriolis acceleration ceases to be an important factor if the tidal acceleration has a much larger magnitude as might be the case, for example, in a narrow river; this would emerge from the dimensional analysis). Pooling this information, we determine the magnitudes of all the terms in the equation. We find that

$$\frac{\partial u}{\partial t} = \qquad u\frac{\partial u}{\partial x} = \qquad fv = \qquad g\frac{\partial \zeta}{\partial x} =$$

$$\frac{\text{friction}}{\rho} =$$

Using $U = 1$, $T = 10^{-4}$ and $L = 10^5$ you will find that *all* of the above terms are of magnitude 10^{-4} except the $u(\partial u/\partial x)$ term (the non-linear term), which is 10^{-5}. Remember that $g = 10$ (approximately) so the magnitude of the term $g(\partial\zeta/\partial x)$ is also 10^{-4}, bearing in mind that the tidal elevation is of the order 1 m. Hence, only this non-linear term can be discarded from our tidal model. It cannot if $L \approx 10^4$ m ($= 10$ km), that is, there are features on the 10 km scale (embayments, headlands, inlets). Again it is recognised that it is important to take non-linear processes into account near coasts where such features predominate.

Example 9.7

A cool sea is adjacent to a warm current, and temperature θ diffuses according to the diffusion equation

$$\frac{\partial \theta}{\partial t} = \kappa \frac{\partial^2 \theta}{\partial x^2},$$

where the rather strange symbols are gradients. $\partial\theta/\partial t$ means the rate of change of temperature with time, and $\partial^2\theta/\partial x^2$ is a rate of change of the temperature gradient ($\partial\theta/\partial x$) with respect to distance (x). Typical scales will be given, since the purpose of this particular example is to run through numerical approximation rather than to give another example of dimensional analysis. Typical lengths are 100 km, times are consistent with the Coriolis parameter, and κ, the diffusion coefficient, has magnitude 4×10^5 m^2 s^{-1}.

Figure 9.8 gives the locations of various measurement sites marked A, B and C. At site A, the gradient of the temperature is 1 °C in 50 km; at site B, which is 100

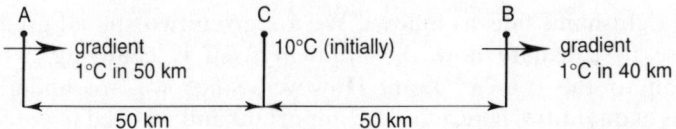

Figure 9.8 Calculating differences from the data.

km east of A, the gradient is 1 °C in 40 km. The directions of these gradients are indicated in Figure 9.8 by the arrows; they are all positive. At site C the temperature is 10 °C. Use the information given to estimate what the temperature will be in 24 hours' time.

The first task is to check that the dimensions of both sides of the diffusion equation are the same and that their magnitude is of the same order. Temperature cannot be expressed in terms of M, L and T (mass, length and time), but since it occurs once on both sides of the diffusion equation, this part at least is in balance. Next, we write down the dimensions of the gradients of θ:

$$\frac{\partial \theta}{\partial t} \text{ has dimensions} = \qquad \frac{\partial^2 \theta}{\partial x^2} \text{ has dimensions} =$$

Your answers should be θT^{-1} and θL^{-2}, respectively. Each time a gradient is required, a derivative is taken which is equivalent to a division by the quantity that varies (the independent variable). Hence, $\kappa(\partial^2 \theta / \partial x^2)$ has dimensions $L^2 T^{-1} \cdot \theta L^{-2} = \theta T^{-1}$, which is the same as the dimensions of $\partial \theta / \partial t$. We are given the dimensions of L and T; these are $L = 10^5$ m and $T = 10^{-4}$ s, and since $\kappa = 4 \times 10^5$ m^2 s^{-1}, we conclude that the two sides of the diffusion equation have the following magnitudes:

$$\frac{\partial \theta}{\partial t} \approx \qquad \kappa \frac{\partial^2 \theta}{\partial x^2} \approx$$

The results of each calculation are $\theta \times 10^{-4}$ for the left-hand side, and $4\theta \times 10^{-4}$ for the right-hand side. This is acceptable bearing in mind that we are only considering orders of magnitude; i.e. we are not saying that 1 is equal to 4, but that 1 and 4 are both in the same ballpark (between 1 and 10, for example). The next part of the question moves us away from such ballpark calculations and forces us to be a little more precise.

The unknown in the last part of the question is the left-hand side, since it is this that will tell us how temperature changes with time. There is enough information to calculate the right-hand side, as follows. We are given two spatial gradients, and we can use these to calculate how the gradient itself is changing, which gives an approximation to the $\partial^2\theta/\partial x^2$ term. However, since we are undertaking actual computations of quantities here, units are important and we need to convert to metres. Doing this, the gradients are as follows:

At A	At B

The answers are, at A, $\frac{1}{5} \times 10^{-4}\theta °C\ m^{-1}$ and at B, $\frac{1}{4} \times 10^{-4}\theta °C\ m^{-1}$. The 'gradient of the gradient' of θ at the point C mid-way between A and B is therefore an estimate of $\partial^2\theta/\partial x^2$; this is

$$\frac{\partial^2\theta}{\partial x^2} \approx $$

Your answer should be $\frac{1}{2} \times 10^{-10}\theta °C\ m^{-2}$ and is obtained from taking the difference between the two values of the gradient and dividing by 100 km, as follows:

$$\frac{\partial^2\theta}{\partial x^2} \approx \frac{\frac{1}{4} \times 10^{-4} - \frac{1}{5} \times 10^{-4}}{10^5} \theta = \frac{1}{2} \times 10^{-10}\theta.$$

Hence $\kappa(\partial^2\theta/\partial x^2) \approx 2 \times 10^{-5}\theta$, and this is also the rate of change of temperature with respect to time, $\partial\theta/\partial t$. Crudely, this can be approximated by the difference

$$\frac{\theta(\text{one day later}) - \theta(\text{now})}{24\ \text{hours}},$$

and since 24 hours is 86 400 or 9×10^4 s, this gives

$$\theta(\text{one day later}) = \theta(\text{now}) + $$

Your answer should be $\theta(\text{now}) + 1.8\,°C$. So the temperature one day later is, roughly, $12\,°C$. Some readers may have found these last few calculations a new experience; for others their crudity may have been all too obvious. They are an improvement on dimensional analysis, but barely so. However, they do provide an introduction to the use of differences as approximations to gradients (derivatives). The calculation of the second-order gradient $\partial^2\theta/\partial x^2$ is in fact similar to the way in which such terms are computed in numerical finite difference schemes. The latter are, as you may guess, far more sophisticated; so much so in fact that the above crude computations are barely recognisable distant cousins. This has enabled us to reach the stage so that we are able to do a systematic example of numerical prediction. As in the first statistics example, oceanographic relevance has had to be sacrificed for clarity.

Example 9.8

This example concerns the prediction of the future population of the United States. At a time t, the population of the USA is given by the function $P(t)$. Let t denote the year past 1900, and the function $P(t)$ is assumed to obey the logistic equation

$$\frac{dP}{dt} = aP - bP^2.$$

We will use the forward difference

$$\frac{dP}{dt} \approx \frac{P(t+h) - P(t)}{h}.$$

The discretised version of the logistic equation using the above forward difference takes the form

$$P(t+h) \approx P(t) + h(aP - bP^2)$$

This is straightforward to obtain from evaluating the logistic equation at time t with a forward difference for dP/dt. It is

$$P(t+h) = P(t) + h(aP(t) + bP^2(t)).$$

With data $a = 0.02$, $b = 4 \times 10^{-5}$, $P(0) = 76.1$ and $h = 10$, calculate $P(10)$.

$P(10) =$

Your answer should be 89.00 (approximately). Continue with $t = 10$ to find $P(20)$ and so on to complete Table 9.5. The answers you should have obtained are contained within Table 9.6. Note how remarkably good the predictions are. One reason for this is that the parameters a and b are both adjustable and they can be chosen so that the fit between actual and predicted population is minimised. If any reader is around in the year 2020, test the formula then to see how good it is! The writer is not very confident.

Table 9.5

Year	t	P(t) (approx.)
1920	20	
1930	30	
1940	40	
1950	50	
1960	60	
1970	70	
1980	80	

Table 9.6

Year	t	P(t) (approx.)	P(t) (measured)
1900	0	76.10	76.1
1910	10	89.00	92.4
1920	20	103.64	106.5
1930	30	120.97	123.1
1940	40	138.21	132.6
1950	50	158.32	152.3
1960	60	179.96	180.7
1970	70	203.00	204.9
1980	80	227.12	226.5

Example 9.9

This example concerns shallow-water equations and numerical approximations to *partial* differential equations. These were encountered in Chapters 3 and 6, albeit superficially.

A simple one-dimensional estuary (or, more strictly, a model that has as its variable a cross-stream average which renders the model one-dimensional) is governed by the equation of motion,

$$\frac{\partial u}{\partial t} + U \frac{\partial u}{\partial x} = -g \frac{\partial \zeta}{\partial x} + ku,$$

where u is the speed of the flow, U is a constant background flow, g is acceleration due to gravity, and ζ is the elevation of the water surface, usually measured from mean sea level (mean low water spring for those with knowledge of tides). The term ku represents in a rather crude way, frictional dissipation. Using only forward differences, put this equation into finite difference form using the notation

$$u_j^s = u(s\Delta t, j\Delta x),$$
$$\zeta_j^s = \zeta(s\Delta t, j\Delta x),$$

such that it predicts a new value of u:

$$u_j^{s+1} = $$

Before giving the answer, there is another relationship which stems from the fact that mass is neither created nor destroyed in the estuary. This takes the rather simple form

$$\frac{\partial \zeta}{\partial t} + h \frac{\partial u}{\partial x} = 0.$$

Here, h is the constant depth of the estuary (the symbol h is universally used for the step length in numerical methods, but we avoid the clash of notation here by retaining the incremental notation Δx and Δt for the space and time steps, respectively). We also write this mass conservation (called continuity) equation in finite difference form:

$$\zeta_j^{s+1} = $$

There are logical reasons for writing the two equations together. They are

$$u_j^{s+1} = -U\left(\frac{\Delta t}{\Delta x}\right)(u_{j+1}^s - u_j^s) - g\left(\frac{\Delta t}{\Delta x}\right)(\zeta_{j+1}^s - \zeta_j^s) + (1 + k\Delta t)u_j^s,$$

and

$$\zeta_j^{s+1} = h\left(\frac{\Delta t}{\Delta x}\right)(u_{j+1}^s - u_j^s) + \zeta_j^s.$$

These two equations are solved in tandem starting from the initial conditions. The main problem with this scheme, which is called *explicit* (because the unknown is on the left of each equation and is explicitly given in terms of known quantities), is that it is unstable for some values of Δt and Δx (the time and space step lengths). Another problem with explicit schemes is their large truncation error, an error due to the poor representation of the gradients (see Chapter 3). As you can see, even this simplified estuarial problem involves solving equations that require a computer. Values for Δt, Δx, U and k are chosen ($g = 9.81$ m s^{-2}, of course), as well as the start values u_0^0 and ζ_0^0. This brings us to a third problem. As posed, what happens at the mouth of the estuary ($x = 0$) dictates what happens upstream. This may not be entirely appropriate; the message here is *always* ask questions as to the physical nature of what is being modelled and see whether the representation of it is right. To program this scheme as it stands would certainly be possible, but the answers would be awry due to the above-mentioned stability, truncation and design problems. An alternative finite difference formulation using centred differences to avoid instability and higher-order terms to improve accuracy would be a far more practical modelling approach. All

Example 9.10

Our last example examines a simple ecosystem model. A three-variable ecosystem model has as variables nutrient (N), phytoplankton (P), and zooplankton (Z), which obey the equations:

$$\frac{dZ}{dt} = b_2 PZ - dZ,$$

$$\frac{dP}{dt} = aNZ - bPZ,$$

$$\frac{dN}{dt} = -aNP + b_1 PZ + dZ,$$

where $b = b_1 + b_2$, a and d are constants, and the units of N, Z and P are milligrams per cubic metre (mg m^{-3}). Time is in days.

By adding these equations together, find a simple relationship between Z, N, and P:

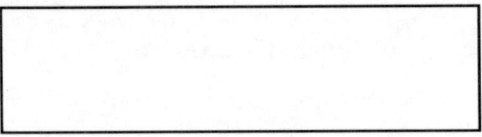

You will find that the right-hand side adds to zero, since if the gradient of a quantity is zero the quantity does not change, which means that the quantity is a constant. Hence $Z + P + N =$ constant. Take this constant as 5.0 (see the paper by Klein and Steele (1985) from which this example is derived). One possible state is $N = 5$, $Z = P = 0$, which is obviously maximum nutrient, but no animals and no plants. This is what is called a steady state solution since it does not change with time. It is also an entirely feasible solution, if a very boring one! Is it the only one? To answer this question, put all the right-hand-side rates of change equal to zero, do not allow any of the variables Z, P or N to be zero, and solve the three simultaneous linear equations:

$$Z = \qquad P = \qquad N =$$

The solutions to these equations are:

$$P = d/b_2,$$

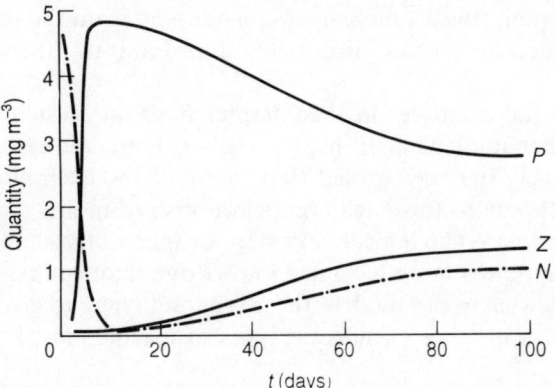

Figure 9.9 The variation of P, Z and N with time. From Klein and Steele (1985). Reproduced with permission of the authors.

$$Z = \left(5 - \frac{d}{b_2}\right) \Big/ \left(1 + \frac{b}{a}\right),$$

$$N = \left(5 - \frac{d}{b_2}\right) \Big/ \left(1 + \frac{a}{b}\right).$$

We are now in a position to experiment with this model a little, although what we can do here is limited by what can be done without running to a computer for help with the solving of the equations. Let us assume that, at time $t = 0$, $N = 5$, $Z = 0$ and $P = 0$. Also assume that, after a long time, the variables Z, P and N reach the above steady state values. First, calculate these given the following values of the constants that appear in the model: $a = 0.2$, $b = 0.15$, $b_2 = 0.03$, and $d = 0.08$.

| Z = | P = | N = |

The arithmetic should reveal the following values:

$Z = 1.33,$ $P = 2.67,$ $N = 1.00.$

However, we are unable to deduce precisely *how* Z, P and N reach these values; this can only be done by solving the equations using for example a marching method (see Example 9.8). The actual shapes of the variations of the three variables are shown in Figure 9.9. P is low at the start, but rapidly climbs to a value close to its maximum (5.0). At this time, both nutrient (N) and zooplankton (Z) values are close to zero. After this time, all variables tend to their predicted eventual steady state values $P = 2.67$, $Z = 1.33$ and $N = 1.0$.

As mentioned earlier, this example has been adapted from the paper by Klein and Steele (1985), in which the role of introducing diffusion into the model is considered in some detail.

It is hoped that the examples in this chapter have, at least in some small way, helped you to understand what is happening when the process of mathematical modelling takes place. It is recognised that many of the examples are so idealised that they are of little use to those who model or wish to model real marine systems. Such real-life modelling is the logical next step for many of you, but to be successful in that endeavour requires more technical knowledge than can be found in this text. For those who only wish to use models, there is enough here to give you some insight into what is behind a number of different types of marine model.

9.4 Exercises

1 Find the mode, median and mean (arithmetic) of the following sets of numbers:
 (a) 3, 5, 7, 5, 1, 9, 5, 4, 8, 2, 3, 7, 4, 9, 5, 1, 6, 2, 3, 1.
 (b) 23, 4, 42, 76, 59, 23, 11, 51, 87, 99, 46, 62, 69, 36, 59, 14, 1, 15, 82, 94.
 (c)

Table 9.7

Height (inches)	Mark (X)	Frequency (f)	fX
60–62	61	5	305
63–65	64	18	1152
66–68	67	42	2814
69–71	70	27	1890
72–74	73	8	584

(d)

Table 9.8

Weight (mg)	Mark (X)	Frequency (f)	fX
0–10	5	4	20
11–20	15.5	11	170.5
21–30	25.5	31	790.5
31–40	35.5	53	1881.5
41–50	45.5	41	1865.5
51–60	55.5	48	2664
61–70	65.5	29	1899.5
71–80	75.5	18	1359
81–90	85.5	8	684

2 Calculate the standard deviation of each of the four sets of data given in Exercise 1.
3 Table 9.9 gives the actual and expected catches of fish over a ten-day period:

Table 9.9

	Species A	Species B	Species C	Species D	Species E
Expected	370	81	46	159	5
Actual	380	79	31	146	6

Using a χ^2 test with $\alpha = 0.01$, determine whether the actual catch deviates significantly from expectation. If $\alpha = 0.75$, which corresponds to a 25% level of significance, is the deviation from expectation significant now? A new set of data is given in Table 9.10.

Table 9.10

	Species A	Species B	Species C	Species D	Species E
Expected	370	81	46	159	5
Actual	369	31	49	148	12

What are the conclusions now? Comment on the interpretation of the statistics according to the χ^2 test as opposed to the interpretation you get from 'eyeballing' the data.
4 The data given in Table 9.11 show the discharges of nitrogen (N) and phosphorus (P) from a river. The total river discharge (Q) is also given. Calculate the correlation coefficients as well as the two regression lines N on Q and P on Q.
5 The distribution of cadmium along the centre line of a polluted river, from the estuary upstream, is presented in two ways: as a table (Table 9.12), and as a graph (Figure 9.10). Calculate the *mean*, *mode* and *median* from the table, and indicate how you would estimate

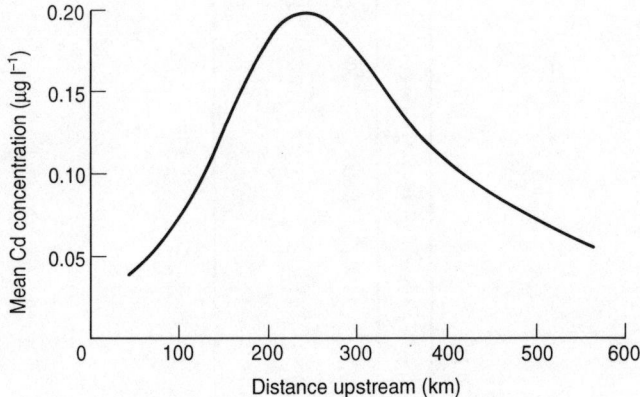

Figure 9.10 Variation of Cd with distance upstream.

Table 9.11

Discharge	1981	1982	1983	1984	1985	1986	1987	1988	1989	1990	1991	1992	1993
N (t/yr)	14 000	8 600	6 400	13 800	12 700	11 200	10 900	16 100	7 900	12 000	13 800	6 000	15 300
P (t/yr)	210	190	140	190	200	160	170	220	120	190	120	90	220
Q (m^3 s^{-1})	120	90	70	80	90	60	70	130	60	60	60	30	130

each of these from Figure 9.10 without the use of Table 9.12. Which of these measures of central tendency gives the clearest indication of the position of the source of the cadmium?

Table 9.12

Distance upstream (km)	Mean Cd concentration ($\mu g\, l^{-1}$)
0–100	0.042
101–200	0.121
201–300	0.195
301–400	0.136
401–500	0.082

How would you expect the variance (as defined in Example 9.1) of the distribution of cadmium to change with time, given that no further cadmium is being discharged into the river?

6 An ocean current is 200 km wide with a speed of 1 m s^{-1} at a latitude of 45°N. If the lateral friction has magnitude 2×10^6 m^2 s^{-1}, calculate the likely dynamic balance. How does this change if the current narrows to 50 km, the current speed increases to 4 m s^{-1} and the friction is now 2×10^5 m^2 s^{-1}?

7 A tidal current has magnitude 5 m s^{-1} in an estuary whose width narrows from 10 to 5 km. The tidal range is 2 m, the period is 12 h and the latitude is 56°N. Friction is known to be important, and if it is equal to $\rho C_D u^2$, where ρ is the density and u is the current speed, what is an appropriate magnitude for C_D?

8 The differential equation

$$\frac{dy}{dt} = 0.1y + 0.02y^2 + t$$

arises from a population model of Antarctic krill. y is the (scaled) biomass and t (of course) is time. By using a forward difference approximation for the left-hand side, with a step size of 0.1, predict the value of y at time $t = 0.5$ given that $y = 1$ when $t = 0$ (this is Euler's method for the solution of differential equations).

Using a more accurate method, y is predicted to be 1.19. Compare your results with this prediction and discuss the reasons for any differences.

9 The differential equation

$$\frac{dx}{dt} = \frac{0.1x}{5t + x}$$

arises from a model that has been built to try and predict the growth of an algal patch which is subject to self-shadowing (x is the patch size, and t is time). Solve this equation as in Exercise 8 by using forward differences (Euler's method), with a step size of 10.0 to predict the size of the patch at time $t = 50$ given $x = 100$ when $t = 0$.

Using a more accurate finite difference method, the patch size is predicted to be 102.5 m. How satisfied are you with your result?

10 The diffusion of salt $S(x, t)$ in a one-dimensional estuary obeys the equation

$$\frac{\partial S}{\partial t} + U \frac{\partial S}{\partial x} = \kappa \frac{\partial^2 S}{\partial x^2},$$

where U is a constant flow, κ is a constant salt diffusion coefficient, x is measured upsteam from the estuary, and t is time. Adopting the notation $S_{i,j} = S(i\Delta t, j\Delta x)$, write this equation in finite difference form using the centred difference form of the first and second derivatives, and forward differences in time. Outline any difficulties you meet in solving this set of equations to predict the salinity along the centre line of the estuary, in particular, the use of boundary values and the likely accuracy of the predictions.

11 The simplest two-dimensional partial differential equation is called Laplace's equation, and can be written

$$\frac{\partial^2 \phi}{\partial x^2} + \frac{\partial^2 \phi}{\partial y^2} = 0.$$

Taking equal steps in the x and y directions, show that the finite difference form of Laplace's equation states that, at any grid point, the value of the variable ϕ is the arithmetic mean of the values of ϕ at the surrounding four grid points. Suppose that this equation is to be solved in an area consisting of 10 000 grid points, what extra information is needed before we can solve the problem? If the finite difference form of Laplace's equation for each of the 10 000 points is written in matrix form,

$$Ax = b,$$

what would be the general form of the matrix A?

Appendix A

Mathematical modelling in oceanography

A.1 Dimensionless numbers

In this appendix, we shall give a more orthodox mathematical treatment of the development of dimensional analysis starting with the equations of motion. It is intended to be understood by those readers with a background roughly equivalent to second-year ancillary mathematics for science degrees or first-year mathematics for engineering in the UK.

The two equations valid in the ocean are a vector equation of motion that arises from Newton's second law applied to the sea in general movement:

$$\frac{\partial \boldsymbol{u}}{\partial t} + (\boldsymbol{u} \cdot \nabla)\boldsymbol{u} + 2\Omega x \boldsymbol{u} = -\frac{1}{\rho} \nabla p + \boldsymbol{g} + \boldsymbol{F},$$

and a scalar equation that expresses the conservation of mass:

$$\nabla \cdot \boldsymbol{u} = 0.$$

These equations are written in vector notation, but in this form they are not easy to non-dimensionalise. They have been written down here simply because the symbols can be defined with brevity as follows: $\boldsymbol{u}$ is the fluid velocity, p is fluid pressure, Ω is the angular velocity of the Earth, ρ is the density (assumed constant), $\boldsymbol{g}$ is apparent gravity, and $\boldsymbol{F}$ represents frictional forces. Axes x, y, z are such that x points east, y points north, and z points directly up; and t is time. The symbol ∇ represents a spatial gradient. In component form, using the accepted eddy viscosity form of friction, these equations become:

$$\frac{\partial u}{\partial t} + u\frac{\partial u}{\partial x} + v\frac{\partial u}{\partial y} + w\frac{\partial u}{\partial z} - fv = -\frac{1}{\rho}\frac{\partial p}{\partial x} + v_H\left(\frac{\partial^2 u}{\partial x^2} + \frac{\partial^2 u}{\partial y^2}\right) + v_V \frac{\partial^2 u}{\partial z^2},$$

$$\frac{\partial v}{\partial t} + u\frac{\partial v}{\partial x} + v\frac{\partial v}{\partial y} + w\frac{\partial v}{\partial z} + fu = -\frac{1}{\rho}\frac{\partial p}{\partial y} + \nu_H\left(\frac{\partial^2 v}{\partial x^2} + \frac{\partial^2 v}{\partial y^2}\right) + \nu_V\frac{\partial^2 v}{\partial z^2},$$

$$\frac{\partial w}{\partial t} + u\frac{\partial w}{\partial x} + v\frac{\partial w}{\partial y} + w\frac{\partial w}{\partial z} = -\frac{1}{\rho}\frac{\partial p}{\partial z} - g + \nu_H\left(\frac{\partial^2 w}{\partial x^2} + \frac{\partial^2 w}{\partial y^2}\right) + \nu_V\frac{\partial^2 w}{\partial z^2},$$

and

$$\frac{\partial u}{\partial x} + \frac{\partial v}{\partial y} + \frac{\partial w}{\partial z} = 0.$$

Defining a typical horizontal length scale L, a typical vertical length scale D, a typical horizontal speed U, a typical time scale T, and a typical Coriolis parameter value as f, these equations can be non-dimensionalised by replacing the variables as follows:

$$x = Lx', \quad y = Ly', \quad z = Dz', \quad t = Tt'$$

$$u = Uu', \quad v = Uv', \quad w = \frac{UD}{L}w',$$

where the last equation for w arises from the continuity equation which implies that

$$\frac{U}{L} = \frac{W}{D}$$

(see Chapter 2). We can now formally substitute for all the unprimed variables that carry a dimension in terms of the primed variables that are dimensionless. The capital letters carry the dimensions. Looking at one equation will give the general idea:

$$\frac{U}{T}\frac{\partial u'}{\partial t'} + \frac{U^2}{L}\left(u'\frac{\partial u'}{\partial x'} + v'\frac{\partial u'}{\partial y'} + w'\frac{\partial u'}{\partial z'}\right) - fUv'$$

$$= -\frac{1}{\rho}\frac{\partial p}{\partial x} + \frac{\nu_H U}{L^2}\left(\frac{\partial^2 u'}{\partial x'^2} + \frac{\partial^2 u'}{\partial y'^2}\right) + \frac{\nu_V U}{D^2}\frac{\partial^2 u'}{\partial z'^2}.$$

Note that in this equation the pressure remains dimensional. The reason for this is that pressure in a fluid adjusts to its surroundings and so the term representing it is not amenable to dimensional analysis. We now divide this equation by fU. This decision is based on the knowledge that it is the Coriolis acceleration that is the dominant term in the momentum balance, and it is with this term that we wish to compare the magnitudes of each of the other terms. The equation has six groups of terms. Ignoring the pressure term, the division gives the following non-dimensional groups that represent their magnitudes: $1/fT$, U/fL, 1, ν_H/fL^2 and ν_V/fD^2. The reader should recognise the second, fourth and fifth expressions as the Rossby number, the horizontal Ekman number and the vertical Ekman number, respectively. Whether you prefer this succinct mathematical derivation here or the more wordy derivation in Chapter 2 is for you to decide.

A.2

An equatorial model

In Section 5.3, a verbal description of the dynamics of the steady circulation in the Pacific equatorial region of the ocean was given. Here we follow this up with a more mathematical description of one specific model. The model chosen is, in fact, not steady but provides a simplified picture of how the main equatorial current depends on the wind and of the characteristic time scales present in the current.

The equatorial current itself takes the form of a narrow jet very close to, but just north of, the equator (see Figure 5.3). The narrowness of the jet may lead one to expect that the dynamics is dominated by the Rossby number; after all, this is the case for that other narrow jet, the Gulf Stream as it crosses the Atlantic. However. there is another important factor. The equatorial current is a straight narrow current, and for such currents the non-linear advective acceleration, $(\boldsymbol{u} \cdot \nabla)\boldsymbol{u}$, is in fact identically equal to zero. This is one of those (fortunately rare) occasions when dimensional analysis is unhelpful, for although the Rossby number may be large, it is multiplied by zero! The model is thus linear, consisting of the other acceleration terms, pressure and gravity forces, and driven by wind stress. The form of the Coriolis parameter used is the so-called equatorial β-plane, $f = \beta y$, where $\beta = 2\Omega/R$, Ω is the angular speed of the Earth and R is the radius of the Earth. Two further points: (1) the ocean is assumed to consist of two layers. This is to model a mixed layer of depth 100 m over a thermocline (modelled as an interface of zero vertical extent) which is in turn above a deep ocean layer that extends 2700 m to the sea bed. This abyssal layer is of course cooler and therefore more dense, but the difference in density is in absolute terms quite small. (2) Because of the two-layer structure of the ocean, the surface wind forcing can be assumed to be a body force acting uniformly through the upper layer. This squares with reality once it is realised that one can integrate through the layer and assume a stress-free thermocline. The effectiveness of gravity is reduced to g', where

$$g' = \frac{\rho_2 - \rho_1}{\rho_1} g,$$

ρ_2 is the density of the lower layer, and ρ_1 is the density of the upper layer. It is well known in the theory of shallow water waves that if the wave speed is c, then $c^2 = gh$, where g is gravity and h is the depth. Here, the wave speed is given by $c^2 = g'H$, where g' is defined above and H is the depth of the mixed layer. This speed is of course much slower than the value for a homogeneous (barotropic) ocean. Some figures will illustrate this:

Ocean depth = 3000 m, $g = 9.81$ m s^{-2},

speed = $\sqrt{3000 \times 9.81} = 171.5$ m s^{-1}.

Mixed layer depth = 100 m, $g' = 0.0196$ m s^{-2},

speed = $\sqrt{100 \times 0.0196} = 1.4$ m s^{-1}.

This much lower speed is the one valid for the two-layer ocean proposed here.

The equations valid under these assumptions are:

$$\frac{\partial u}{\partial t} - fv = -g' \frac{\partial \zeta}{\partial x} + \frac{\tau^x}{H},$$

$$\frac{\partial v}{\partial t} + fu = -g' \frac{\partial \zeta}{\partial y} + \frac{\tau^y}{H},$$

$$g' \frac{\partial \zeta}{\partial t} + c^2 \left(\frac{\partial u}{\partial x} + \frac{\partial v}{\partial y} \right) = 0.$$

The first two equations are recognisably the x and y equations of motion with the pressure gradient replaced by a sea surface slope. This can be done because pressure is in hydrostatic balance, so the pressure at a given point is dependent only on the weight of water above that point. Gradients in pressure will therefore depend solely upon the change of this weight with horizontal position, and this is in turn solely dependent on the sea surface slope. Also, we have already mentioned, the body force composed of the surface wind stress averaged over the mixed layer. The notation used here is τ^x for eastward wind stress, and τ^y for northward wind stress. The third equation is harder to recognise. It is a vertically integrated continuity equation, the $\partial \zeta / \partial t$ term arising from integrating $\partial w / \partial z$ with respect to z, and equating w with $\partial \zeta / \partial t$ at the sea surface. These equations can be solved (three equations in three unknowns). The details of how these equations are solved are themselves dependent on what the modeller wishes to extract in the way of information. To solve these numerically using a finite difference scheme is entirely possible, although our present purposes are best served if the variables ζ and u are formally eliminated to leave a single equation in v. Further, since we are interested in the equatorial jet where variations in the x direction are much more gradual than variations in y direction or t (time), we can assume $v = v(x, t)$. This simplifies the equation obeyed by v into:

$$\frac{\partial^2 v}{\partial t^2} + f^2 v - c^2 \frac{\partial^2 v}{\partial y^2} = -\frac{f \tau^x}{H}.$$

The dominant balance here is $v = -\tau^x/(fH)$ obtained by ignoring the second derivative terms. This is the expression obtained from pure Ekman balance. Recall Chapter 2 where Ekman balance was due to the Coriolis acceleration matching the dissipation due to the vertical transfer of momentum (in simple terms, the vertical eddy viscosity). Integrating this balance vertically through the Ekman layer gives the above expression for v. As we are very near the equator, it is perhaps surprising that the Coriolis acceleration still plays a prominent role, especially since at the equator itself f is zero, which would render v locally infinite. In fact, the actual jet is just north of the equator, and, as the equator is approached, the as-yet ignored derivative terms in the above equation for v become significant. To be more quantitative, the Ekman balance is valid provided the distance of the jet from the equator exceeds the value c/f, which is called the local radius of deformation (a rather odd phrase). More details of this model are given in Philander (1990), but two interesting scales that can be derived are the distance $\sqrt{c/\beta}$, called the equatorial radius of deformation, and the time scale $1/\sqrt{\beta c}$, which determines the relative importance of the first two terms of the v equation. The equatorial radius of deformation has a value of about 250 km, and is the distance from the equator within which the second and third terms of the v equation are of comparable magnitude. The time scale $1/\sqrt{\beta c}$ has the value 1.5 days and is a local inertial time (usually $1/f$) at latitude $\sqrt{c/\beta}$.

For those that can stand the pace of all this mathematics, what has been presented here are two examples that display some of the power of dimensional analysis and some of the understanding that can emerge from careful modelling as distinct from solving a full set of equations using a large computer code.

A.3 Modelling *El Niño*

It would provide a very fitting close to this short appendix on the more mathematical side of modelling if a simple model that predicts *El Niño* could be given. Unfortunately this is not possible at the moment because the details of the underlying cause of *El Niño* are not known to a sufficient degree. However, the basic cause of the southern oscillation is known, and the mathematical details behind the basic model are now given. The main ingredient necessary for the prediction of the southern oscillation that is absent from the previous equatorial model is a horizontal temperature gradient. It also turns out that it is essential to include the variation of the mean easterly flow with latitude, $U(y)$. This may seem rather an obscure feature to include in the simplest model, but it is this variation that ensures that vorticity is conserved. Recall from Chapter 5 that the further away from the equator, the greater is the planetary vorticity. (Vorticity as a whole is conserved. This means that the sum of the two contributions from the variation of the Coriolis parameter with latitude, the *planetary* vorticity, and from the contribution from the dynamics of the current itself, the *fluid* vorticity, must be zero.) Thus, the further away from the equator, the less is the fluid vorticity, to compensate for the increase in planetary vorticity. In the following simple model, there is only a zonal flow. Therefore, this decrease in fluid vorticity can only arise from a change in the magnitude of the flow with latitude, hence $U = U(y)$. For those with a working knowledge of fluid mechanics, Couette flow (the flow close to a plane surface) exhibits a similar vorticity. The complete set of equations that includes the above effects is as follows:

$$\frac{\partial u}{\partial t} + U \frac{\partial u}{\partial x} - \left(f - \frac{\partial U}{\partial y}\right)v = -g' \frac{\partial \zeta}{\partial x},$$

$$\frac{\partial v}{\partial t} + U \frac{\partial v}{\partial x} + fu = -g' \frac{\partial \zeta}{\partial y},$$

$$\frac{\partial \zeta}{\partial t} + U \frac{\partial \zeta}{\partial x} + \frac{\partial}{\partial x}(Hu) + \frac{\partial}{\partial y}(Hv) = 0,$$

where $H(x, y)$ is the depth of the thermocline. This set of equations describes linear wave motion within the equatorial regions of the ocean. The thermodynamics, said at the outset to be central to the prediction of *El Niño* are hard to detect in this model, but they are there. The changing depth of the thermocline $H(x, y)$ contains adequate thermodynamics, since the deeper the thermocline, the greater the amount of heat there is in that part of the ocean. These equations differ from the previous set only in that they explicitly include the mean flow $U(y)$ and its all important northwards derivative $\partial U/\partial y$. These equations then provide the basis of the analysis of the southern oscillation and hence provide the first steps towards understanding the phenomenon that is *El Niño*. Even further linear solutions of this set of equations are beyond the scope of this text; see Philander (1990) for an account of them. Actually solving them with real input data representative of the equatorial Pacific requires the setting up of a numerical model. As has already been said, even this does not adequately model *El Niño*. It cannot, for example, predict at what years there are likely to be *El Niño* events because the solutions of these linear equations will lead to periodic (wave-like) behaviour of the variables. It is precisely the lack of a predictable periodicity that characterises *El Niño*. The above is, however, a first step on what it is unfortunately likely to be a very long road.

Appendix B
The χ^2 distribution

Percentile values (χ_p^2) for the χ^2 distribution, with v degrees of freedom

Levels of significance

v	$\chi^2_{0.995}$	$\chi^2_{0.99}$	$\chi^2_{0.975}$	$\chi^2_{0.95}$	$\chi^2_{0.90}$	$\chi^2_{0.75}$	$\chi^2_{0.50}$	$\chi^2_{0.25}$	$\chi^2_{0.10}$	$\chi^2_{0.05}$	$\chi^2_{0.025}$	$\chi^2_{0.01}$	$\chi^2_{0.005}$
1	7.88	6.63	5.02	3.84	2.71	1.32	0.455	0.102	0.0158	0.0039	0.0010	0.0002	0.0000
2	10.6	9.21	7.38	5.99	4.61	2.77	1.39	0.575	0.211	0.103	0.0506	0.0201	0.0100
3	12.8	11.3	9.35	7.81	6.25	4.11	2.37	1.21	0.584	0.352	0.216	0.115	0.072
4	14.9	13.3	11.1	9.49	7.78	5.39	3.36	1.92	1.06	0.711	0.484	0.297	0.207
5	16.7	15.1	12.8	11.1	9.24	6.63	4.35	2.67	1.61	1.15	0.831	0.554	0.412
6	18.5	16.8	14.4	12.6	10.6	7.84	5.35	3.45	2.20	1.64	1.24	0.872	0.676
7	20.3	18.5	16.0	14.1	12.0	9.04	6.35	4.25	2.83	2.17	1.69	1.24	0.989
8	22.0	20.1	17.5	15.5	13.4	10.2	7.34	5.07	3.49	2.73	2.18	1.65	1.34
9	23.6	21.7	19.0	16.9	14.7	11.4	8.34	5.90	4.17	3.33	2.70	2.09	1.73
10	25.2	23.2	20.5	18.3	16.0	12.5	9.34	6.74	4.87	3.94	3.25	2.56	2.16
11	26.8	24.7	21.9	19.7	17.3	13.7	10.3	7.58	5.58	4.57	3.82	3.05	2.60
12	28.3	26.2	23.3	21.0	18.5	14.8	11.3	8.44	6.30	5.23	4.40	3.57	3.07
13	29.8	27.7	24.7	22.4	19.8	16.0	12.3	9.30	7.04	5.89	5.01	4.11	3.57
14	31.3	29.1	26.1	23.7	21.1	17.1	13.3	10.2	7.79	6.57	5.63	4.66	4.07
15	32.8	30.6	27.5	25.0	22.3	18.2	14.3	11.0	8.55	7.26	6.26	5.23	4.60
16	34.3	32.0	28.8	26.3	23.5	19.4	15.3	11.9	9.31	7.96	6.91	5.81	5.14
17	35.7	33.4	30.2	27.6	24.8	20.5	16.3	12.8	10.1	8.67	7.56	6.41	5.70
18	37.2	34.8	31.5	28.9	26.0	21.6	17.3	13.7	10.9	9.39	8.23	7.01	6.26
19	38.6	36.2	32.9	30.1	27.2	22.7	18.3	14.6	11.7	10.1	8.91	7.63	6.84
20	40.0	37.6	34.2	31.4	28.4	23.8	19.3	15.5	12.4	10.9	9.59	8.26	7.43
21	41.4	38.9	35.5	32.7	29.6	24.9	20.3	16.3	13.2	11.6	10.3	8.90	8.03
22	42.8	40.3	36.8	33.9	30.8	26.0	21.3	17.2	14.0	12.3	11.0	9.54	8.64
23	44.2	41.6	38.1	35.2	32.0	27.1	22.3	18.1	14.8	13.1	11.7	10.2	9.26
24	45.6	43.0	39.4	36.4	33.2	28.2	23.3	19.0	15.7	13.8	12.4	10.9	9.89
25	46.9	44.3	40.6	37.7	34.4	29.3	24.3	19.9	16.5	14.6	13.1	11.5	10.5
26	48.3	45.6	41.9	38.9	35.6	30.4	25.3	20.8	17.3	15.4	13.8	12.2	11.2
27	49.6	47.0	43.2	40.1	36.7	31.5	26.3	21.7	18.1	16.2	14.6	12.9	11.8
28	51.0	48.3	44.5	41.3	37.9	32.6	27.3	22.7	18.9	16.9	15.3	13.6	12.5
29	52.3	49.6	45.7	42.6	39.1	33.7	28.3	23.6	19.8	17.7	16.0	14.3	13.1
30	53.7	50.9	47.0	43.8	40.3	34.8	29.3	24.5	20.6	18.5	16.8	15.0	13.8
40	66.8	63.7	59.3	55.8	51.8	45.6	39.3	33.7	29.1	26.5	24.4	22.2	20.7
50	79.5	76.2	71.4	67.5	63.2	56.3	49.3	42.9	37.7	34.8	32.4	29.7	28.0
60	92.0	88.4	83.3	79.1	74.4	67.0	59.3	52.3	46.5	43.2	40.5	37.5	35.5
70	104.2	100.4	95.0	90.5	85.5	77.6	69.3	61.7	55.3	51.7	48.8	45.4	43.3
80	116.3	112.3	106.6	101.9	96.6	88.1	79.3	71.1	64.3	60.4	57.2	53.5	51.2
90	128.3	124.1	118.1	113.1	107.6	98.6	89.3	80.6	73.3	69.1	65.6	61.8	59.2
100	140.2	135.8	129.6	124.3	118.5	109.1	99.3	90.1	82.4	77.9	74.2	70.1	67.3

Source: Catherine M. Thompson, Table of percentage points of the χ^2 distribution, *Biometrika* **32** (1941), by permission of the author and publisher.

Appendix C

Commercially available software

C.1 Commissioned software

When an industrial establishment wishes to model a particular stretch of sea, the commonest course of action to take, even now, is for the company to commission an expert to build a model to the required specification. The company is usually an engineering firm that is too small to possess its own modelling section or equally well it may be a large company but new to sea modelling. The requirements of such corporations usually fall into one or more of the following categories:

1. A detailed model of the currents in a particular coastal region or embayment, usually as an aid to the construction of a marina or other waterside facility.
2. A water quality model, usually covering a river and estuary system. There is a requirement to satisfy environmental legislation set down by government bodies, and a working model forms a crucial part of the case submitted for compliance with such legislation. There are obvious public relations bonuses to be reaped from a reliably safe prediction of a model in this environmentally sensitive age.
3. An offshore model of a restricted area including both currents and waves. This is usually for a company wishing to build an installation offshore for the extraction of hydrocarbons or other minerals.
4. Ministry of Defence or military establishments sometimes require ocean models. These models often need to be both extensive and detailed.

All of these customers are reasonably wealthy and are able to afford to commission a model for their own exclusive use. The cost of this is usually about £20 000 in 1995 prices. This purchasing strategy, unfortunately, bars the software from the open market (especially in the case of 4). In older times, it was university academics who almost exclusively built this software under consultancy contracts. Nowadays it is more often than not a software company or a newly privatised government research establishment or QUANGO (quasi-autonomous national government organisation).

142   Appendix C

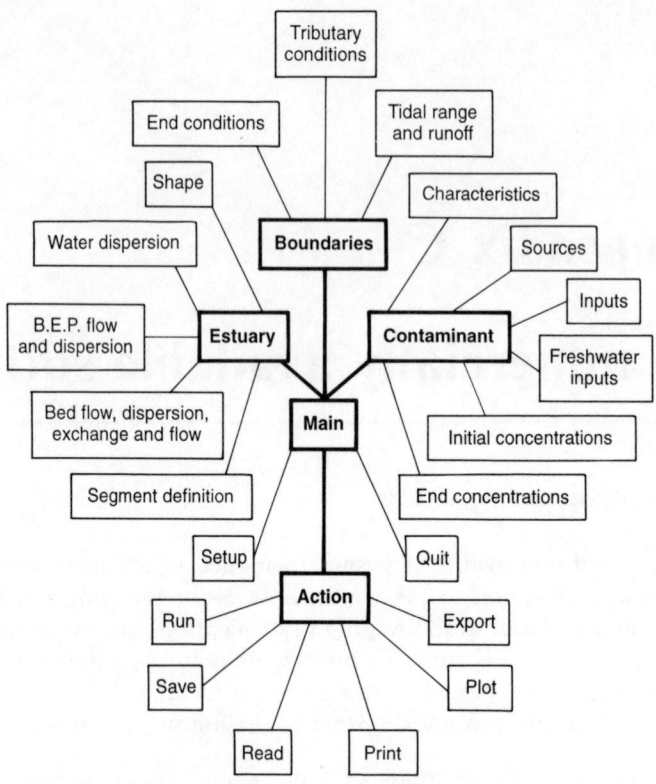

Figure C.1 The ECoS menu system. The central menus provide access to further menus as shown, and can be directly accessed from any of these menus.

The research establishments that have the best track record for producing such software in Europe are: Delft Hydraulics in The Netherlands, Proudman Oceanographic Laboratory in Merseyside, UK, Hydraulics Research in Wallingford, UK, Sir William Halcrow and Partners in Swindon, UK, and Oceanor in Trondheim, Norway. Universities are still producing relevant software; perhaps the best UK source is Professor R.A. Falconer at Bradford University. Other large companies have their own in-house modelling sections; for example, Zeneca has an Environmental Modelling Group that serves the needs of Zeneca (and ICI, from which it sprang), and also tenders for commercial work. Finally, the agencies responsible for water quality, the NRA (National Rivers Authority) in England and Wales, the River Purification Boards in Scotland, and the Water Authorities, have their own modellers but also commission models from others from time to time. As the older models are superseded, the customers who paid for them tend to become less sensitive about releasing the details into the public domain.

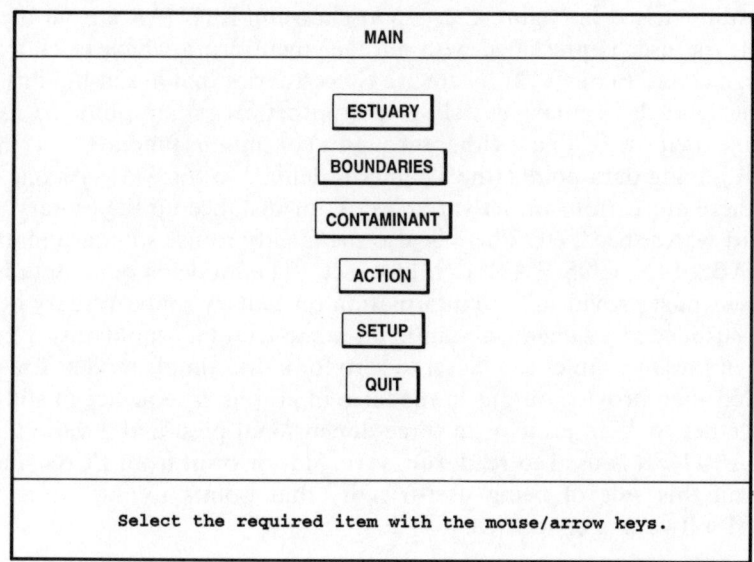

Figure C.2 The first ECoS menu.

One of the first models to emerge from Delft Hydraulics was a depth-averaged two-dimensional model of the southern North Sea. This model is now freely available, at least to the academic research community, through the Proudman Oceanographic Laboratory. One can take it and modify the code (it's in FORTRAN) to include different sea bed boundary conditions, sediment dynamics or whatever. This kind of thing is likely to become more common. To finish off this short appendix, an account is given of a commercially available system shell.

C.2 The ECoS system

ECoS (Estuarial Contaminant System) is a piece of commercial software that retails at several hundred pounds sterling and comes in the form of three floppy disks and a manual. The software will run on any PC, even a 286, but the latest version demands a Windows environment which is, as a minimum, version 3.1 and hence the microcomputer needs to be at least a 386, preferably with a maths coprocessor chip. Enough of all this computer jargon! What follows is a brief run through of the operation of ECoS version 2 (as I write, version 3 is due for release).

ECoS is not a model, but a modelling environment. When the software is run (or 'booted up' in the jargon) the user is presented with a menu in the form of a series of boxes (see Figure C.1). The user selects the box of his or her choice using the mouse or the cursor keys. The entire system shell is set out in a similar fashion; the menu system is summarised in Figure C.2. Upon choosing ESTUARY from the menu in Figure C.1, the user is presented with another menu from which to choose (shape, dispersion, etc.; see Figure C.2). There are two estuaries that are in-built as examples (the Thames and the Tamar) as well as a hypothetical estuary, but the user is also free to define their own estuary, either through a continuous function (as is the Tamar) or through joining data points (the Thames is defined in this way). As one builds up a model, there are various preset variables: X for distance up the estuary, U for the net seaward water speed, etc. There is also the facility to use standard mathematical functions ABS, SIN, COS, TAN, LN, LOG, etc. The model is quite data hungry, in that the user must provide all the information on estuary shape, estuary depth, tidal dynamics, suspended sediment, contaminant input, tributary input, dissolved oxygen, etc. Many of these variables can be set to zero for a first simple model. The programs within ECoS then provide output in the form of graphs, a sequence of stills that can be run together to form a movie, a three-dimensional plot, and a variety of others. The term ACTION is used to read, run, save, plot or print from ECoS, and it is the emphasis on this side of being user-friendly that points to the future as far as commercial software is concerned.

Answers to exercises

1 (a) 4.5, (b) 47.65, (c) 67.45, (d) 46.64.
2 (a) 2.565, (b) 31.00, (c) 2.935, (d) 18.096.
3 $\chi^2 = 2.092$. Not significant at the 99% level; significant at the 25% level. New table, $\chi^2 = 41.62$. Significant at both levels due to outliers in the data.
4 Corr. coefficient = 0.7281   regr. line is $y = 65.55 + 0.0092x$
   Corr. coefficient = 0.708   regr. line is $y = 6.537 + 0.0065x$
5 Mean = 267,* mode = 250, median = 264.
   The variance would increase, theoretically proportionally to time. (*This is the mean *of the distribution*, not the mean concentration, which is 0.1152, the average concentration over the estuary.)
6 Horizontal friction versus Coriolis (large horizontal Ekman number, small Rossby number).
   New figures, advection versus Coriolis (small horizontal Ekman number, large Rossby number).
7 10^9 in SI units.
8 1.16 (Euler). Truncation error.
9 102.92, reasonably satisfied.
10 $S_{i+1,j} = S_{i,j} - \dfrac{U\Delta t}{2\Delta x}(S_{i,j+1} - S_{i,j-1}) + \dfrac{\kappa \Delta t}{(\Delta x)^2}(S_{i,j-1} - 2S_{i,j} + S_{i,j-1})$.

11 Boundary conditions.

$$A = \begin{pmatrix} -4 & 1 & 0 & 0 & \cdots \\ 1 & -4 & 1 & 0 & \cdots \\ 0 & 1 & -4 & 1 & \cdots \\ 0 & 0 & 1 & -4 & \cdots \\ \vdots & \vdots & \vdots & \vdots & \end{pmatrix}.$$ A banded matrix.

References

Backhaus, J.O. (1983) A semi-implicit scheme for the shallow water equations for application to shelf sea modelling, *Cont. Shelf Res.* **2**(4): 243–54.

Banks, J.E. (1974) A mathematical model of a river–shallow sea system used to investigate tide, surge and their interaction in the Thames–southern North Sea region *Phil. Trans. R. Soc. A* **275**: 567–609.

Clements, R.R. (1989) *Mathematical Modelling: A Case Study Approach* (Cambridge: Cambridge University Press).

Cross, M. and Moscardini, A.O. (1985) *Learning the Art of Mathematical Modelling* (Chichester: Ellis Horwood).

Davies, A.M. (1987) Numerical modelling of marine systems in *Numerical Modelling: Applications to Marine Systems*, Ed. J. Noye (Amsterdam: Elsevier) pp. 1–24.

Dyke, P.P.G. (1980) On the Stokes' drift induced by tidal motions in a wide estuary, *Est. Coastal Mar. Sci.* **11**: 17–25.

Dyke, P.P.G. (1987) Water circulation in the Firth of Forth, Scotland, *Proc. R. Soc. Edin.* **93B**: 273–84.

Dyke, P.P.G., Moscardini, A.O. and Robson, E.H. (Eds) (1985) *Offshore and Coastal Modelling*, Lecture notes in Coastal and Estuarine Studies 12 (Berlin: Springer).

Dyke, P.P.G. and Robertson, T. (1985) The simulation of offshore turbulence using seeded eddies, *App. Math. Mod.* **9**(6): 429–33.

Edwards, D. and Hamson, M. (1989) *A Guide to Mathematical Modelling* (Basingstoke: Macmillan).

Ekman, V.W. (1905) On the influence of the Earth's rotation on ocean currents *Arkiv. für Matematik, Astronomi och Fysik*, **2**(11): 52.

Finlayson, B.A. (1972) *The Method of Weighted Residuals and Variational Principles* (London: Academic Press).

Flather, R.A. (1979) in *Marine Forecasting*, Ed. J.C.J. Nihoul (Amsterdam: Elsevier) pp. 385–409.

Goldberg, E.D. *et al.* (Eds) (1977) *The Sea: Ideas and Observations*, Vol. 6 *Marine Modelling* (New York: Wiley–Interscience).

Heaps, N.S. (1987) *Three-Dimensional Coastal Ocean Models*, Coastal and Estuarine Sciences 4. American Geophysical Union.

Hunter, J.R. (1980) An interactive computer model of oil slick motion, *Oceanol. Int. Session M 42–50*, Brighton, UK.

Jeffers, J.N.R. (1988) *Practitioner's Handbook on the Modelling of Dynamic Change in Ecosystems*, SCOPE 34 (Chichester: Wiley).

Jørgensen, S.E. (1986) *Fundamentals of Ecological Modelling* (Amsterdam: Elsevier).
Klein, P. and Steele, J.H. (1985) Some physical factors affecting ecosystems, *J. Marine Res.* **43**(2): 337–43.
Leblond, P.H. and Mysak, L.A. (1978) *Waves in the Ocean* (Amsterdam: Elsevier).
Lighthill, M.J. (1968) Dynamic response of the Indian Ocean to onset of the Southwest Monsoon, *Phil. Trans. R. Soc.* **265A**: 45–92.
Mellor, G.L. and Yamada, T. (1974) A hierarchy of turbulence closure models for planetary boundary layers, *J. Atmos. Sci.* **31**: 1791–896.
Mitchell, A.R. and Wait, R. (1985) *Finite Element Analysis and Applications* (Chichester: Wiley).
Munk, W.H. (1950) On the wind-driven ocean circulation, *J. Meteorol.* **7**: 79–93.
Murdoch, J. and Barnes, J.A. (1974) *Statistical Tables* (London and Basingstoke: Macmillan).
Okubo, A. (1971) Oceanic diffusion diagrams, *Deep Sea Res.* **18**: 781–802.
Pedlosky, J. (1987) *Geophysical Fluid Dynamics*, 2nd edn (Berlin: Springer).
Philander, S.G. (1990) *El Niño, La Niña, and the Southern Oscillation* (London: Academic).
Pickard, G.L. and Emery, W.J. (1982) *Descriptive Physical Oceanography: An Introduction*, 4th edn (Oxford: Pergamon).
Pond, S. and Pickard, G.L. (1991) *Introductory Dynamical Oceanography*, 2nd edn (Oxford: Pergamon).
Radach, G. (1983) Simulation of phytoplankton dynamics and their interactions with other system components during FLEX '76 in *North Sea Dynamics*, Eds. Sundermann and Lenz (Berlin: Springer) pp. 584–610.
Rassmussen, E.H. and Carpenter, T.H. (1982) Variations in tropical sea surface temperature and surface wind fields associated with the southern oscillation/El Niño, *Month. Weath. Rev.* **110**: 354–84.
Ropelowski, C.F. and Halpert, M.S. (1987) Global and regional scale precipitation patterns associated with the El Niño/Southern Oscillation, *Month. Weath. Rev.* **115**: 1606–26.
Stommel, H. (1948) The westward intensification of wind-driven currents, *Trans. Am. Geophys. Union* **29**: 202–6.
Stommel, H. (1965) *The Gulf Stream*, 2nd edn (Berkeley, CA: University of California Press).
Sverdrup, H.U. and Munk, W.H. (1947) Wind, sea and swell: theory of relations for forecasting, *US Hydrogr. Office Publication* 601, p. 44.
Taylor, A.H., Watson, A.J., Ainsworth, M., Robertson, J.E. and Turner, D.R. (1991) A modelling investigation of the role of phytoplankton in the surface balance of carbon at the surface of the North Atlantic, *Global Biogeochemical Cycles* **5**(2): 151–71.
Taylor, A.H., Watson, A.J. and Robertson, J.E. (1992) The influence of the spring phytoplankton bloom on carbon dioxide and oxygen concentrations in the surface waters of the northeast Atlantic during 1989, *Deep Sea Res.* **39**(2): 137–52.
Thompson, C.M. (1941) Table of percentage points of the χ^2 distribution, *Biometrika*, **32**.
Varela, R.A., Cruzado, A., Tintoré, J. and Ladona, E.G. (1994) Modelling the deep-chlorophyll maximum: A coupled physical–biological approach, *J. Marine Res* **50**(3): 441–63.
Webb, A.J. and Metcalfe, A.P. (1987) Physical aspects, water movements and modelling studies of the Forth Estuary, *Proc. R. Soc. Edin.* **93B**: 259–72.
Westerink, J.J., Luettich, R.A. and Muccino, J.C. (1994) Modeling tides in the western North Atlantic using unstructured graded grids, *Tellus* **46A**(2): 178–99.
Williams, J. and Elder, S.A. (1989) *Fluid Physics for Oceanographers and Physicists* (Oxford: Pergamon).

Index

acceleration, advective   8, 9, 11
acceleration, point   8
amphidromic points   41
Antarctic circumpolar current   52, 54
Arakawa A grid   27
Arakawa B grid   27
Arakawa C grid   27
aspect ratio   12
average (see mean)   104

backward difference   19
baroclinic   46, 75
barotropic   46, 75
beta-plane (or effect)   50, 56, 135
biochemical effects   92
bottom friction   16
boundary conditions   29, 30, 32–36, 71
boundary layer   32, 34
boundary, western   43, 44
box models   66, 67
Brunt-Väisälä frequency   13, 47, 48, 98
Buckingham Pi theorem   16
buoyancy frequency   13, 47, 48, 98

centred difference   19
chi-squared test   37, 109, 110, 111
Chlorophyll-alpha   96
coastal boundary   34
coastal upwelling   49
concentration gradient   79
continental shelf model   20, 60–76
continental slope   60

corange line   63
Coriolis acceleration   7
Coriolis effect   11
Coriolis parameter   14, 48, 118, 118, 120
correlation   112, 115, 117
correlation, Pearson   114
cotidal line   63
countercurrent   46
current,
   Antarctic circumpolar   52, 54
   boundary   45, 46
   counter   46
   Ekman   65
   equatorial   50, 135
   geostrophic   42
   Somali   49
   surface   65
   tidal   63
   western boundary   44
   wind driven   65

data assimilation   30, 31
density   8, 13, 66
density gradient   13
depth of friction influence (see Ekman layer)   65
derivative   18
diffusion   77, 95, 97
diffusion coefficient   120
dimension   8
dimensional analysis   6, 133
dimensionless groups   11

149

discretization   3
dispersion   77
Dooley current   85
drag coefficient   17
drying (in models)   35

ECoS   143
ecosystem   88
ecosystems models   89
eddies   10
eddy viscosity   10, 32, 68, 78, 98
Ekman effect (balance)   47, 82
Ekman layer   65
Ekman number   16, 134
Ekman number, horizontal   11, 134
Ekman number, vertical   11, 68, 134
Ekman suction   65
El Niño   50, 51, 56–59, 137–138
equation,
   difference   24
   hydrostatic   120
   of motion   6, 133
   of state   12, 98
equatorial countercurrent   50, 51
equatorial current   50, 51, 135
equilibrium point   91
error   21
error, round-off   22
error, truncation   22, 23
estuarial flow   66
estuary   15, 73
Euler's method   123, 131
evaporation   32

Fickean diffusion   78, 79, 81, 83
fine resolution   52
Fine Resolution Antarctic Model
    (FRAM) 52, 53, 54
finite differences   19, 38, 125
finite element method   21, 25, 40
Firth of Forth   73
flow,
   geostrophic   42
   mean   61

   tidal   63
   turbulent   12, 32
force   9, 10
forward difference   19, 24
frequency polygon   105
friction, bottom   16
Froude number   15, 69
fundamental quantities   6

geostrophic balance   42
global ocean modelling   43
gradients   121
gravity   11
gravity, reduced   14, 15
grid   19, 20, 23, 26, 27
Gulf Stream   14, 15, 16, 46, 48
Gulf Stream rings   48
gyres   43

halocline   13
heterotrophs   99
high-level programming language   2
hydrostatic balance   55, 61, 136

inference   107
internal waves   13, 15
interpolation   37
irradiance   93

Kelvin wave   62, 63, 71
Kolmogorov law   83

La Niña   58
Langmuir circulations   63, 67, 69, 71, 78
large scale modelling   52
layer,
   boundary   32, 34
   Ekman   65
   mixed   46, 65, 135
length   6
level of significance   110
linear friction law   34
log–linear plot   112
log–log plot   112

logarithm   112, 113
logarithmic layer   33
logistic equation   88, 123
logistic growth   88, 123
Lotkas-Volterra model   97

mass   6
mean   104
mean sea level   61
median   104
Michaelis–Menton relationship   93, 95
mixed layer   46, 65, 135
Modal method   21, 74
mode   74, 104
model, numerical   3, 18
model resolution   24
modelling   1
modelling, art of   1
monsoon   51
Monte-Carlo technique   82

nested models   36
Newton's second law   7
non-linear effects (see Rossby number)   120
no-slip condition   34, 46
null hypothesis   37, 109, 111
number,
   Ekman   16, 134
   Froude   15
   Prandtl   12, 48
   Reynolds   12
   Richardson   11, 13, 98
   Rossby   11, 14, 16, 51, 118, 134, 135
   Taylor   12
numerical methods   3, 18
numerical tidal model   70
nutrient   126

ocean,
   Antarctic   54
   Atlantic   48
   modelling   52, 53
   Pacific   50

oil slick, surface   82
oil spillage   81
open boundary   35, 71

parameter, Coriolis   14, 48, 118, 119, 120
parameters   5
particle tracking   78, 81, 87
phytoplankton   95, 126
planetary vorticity   45, 137
pollutant   85
pollution   77
population studies   89, 90
Prandtl number   12, 48
predator-prey   89
pressure force   10, 11
pressure gradient   10
pycnocline   13

quadratic friction law   17, 65, 71, 72, 119

rate equations   100
reduced gravity   14, 15
regression line   79, 112, 116, 117
Reynolds number   12
Richardson number   11, 13, 98
Richardson number, gradient   11, 47
rigid-lid approximation   30, 31
Rossby number   11, 14, 16, 51, 118, 134, 135
Rossby radius (of convergence, deformation)   14, 15, 51, 69, 75
Rossby radius (internal)   15, 75
rotation of earth   7
roughness elements   33

scaling factor   1
scatter diagram   112
scheme, implicit   25
scheme, explicit   25
sea bed boundary   32
sea surface boundary   31
sediment   85
semi-implicit (finite difference scheme)   23

Semtner model   54–56
sensitivity   96
shear, current   45, 47
shear, velocity   47
sigma coordinates   28, 53
slip velocity   34
Somali current   49, 51
spectral method   21
spin-up (of a model)   38
spreading   87
stability   22
staggered grid   23
standard deviation   107
statistics   103
steady ocean circulation models   45
stochastic models   90
Stokes' drift   68
storm surge   71
stress   32, 68
stratification   13
swell waves   9
systems approach   4

t-distribution   114
Taylor number   12
temperature gradient   31
thermocline   13
three-dimensional grid   21
tidal current   63
tidal modelling   60, 71
time   6

time splitting technique   75
trophic model   101
turbulence   10
turbulence closure   98
turbulence closure scheme   32, 33

undercurrent   50
units   8
upwelling   12, 49

validation   2, 37, 41
variable, non-dimensional   134
variables   5
variance   79, 107
viscosity, dynamic   8
viscocity, kinematic   8, 10
vorticies   67
vorticity   45
vorticity, fluid   45, 137

wave drift   68
wave slope   61, 68
western intensification   43, 44
wind-driven circulation   65, 67
wind stress   31
World Ocean Circulation Experiment
   (WOCE)   52, 53, 54

z-statistic   109
zone, equatorial   134–137
zooplankton   101, 126